# The Early Medieval Pottery Industry at al-Basra, Morocco

Nancy L. Benco

BAR International Series 341
1987

# B.A.R.

5, Centremead, Osney Mead, Oxford OX2 0DQ, England.



B.A.R. -S341, 1987: 'The Early Medieval Pottery Industry at al-Basra, Morocco'



ISBN 9780860544401 paperback
ISBN 9781407346144 e-book
DOI https://doi.org/10.30861/9780860544401
A catalogue record for this book is available from the British Library
This book is available at www.barpublishing.com

TABLE OF CONTENTS

LIST OF ILLUSTRATIONS

## LIST OF TABLES

## PREFACE

This monograph reports the results of field work conducted in 1980 and 1981 at the early Islamic site of al-Basra (ca. A.D. 800-1100) in northern Morocco, under the auspices of the American-Moroccan Archeological Mission directed by Dr. Charles L. Redman. It also presents the results of a detailed study of the al-Basra pottery undertaken by the author in 1983 in Tangier and subsequently at the Department of Anthropology, State University of New York at Binghamton, and the Conservation Analytical Laboratory, Smithsonian Institution, Washington, D.C. This publication is a revised version of my Ph.D. dissertation submitted in May 1986.

Broadly speaking, the excavations at al-Basra were undertaken to investigate a number of issues related to the development of cities and states in North Africa, among them the interaction between inland capitals and coastal entrêpots. Two seasons of field work at the site have begun to address some of these questions. The excavations were also undertaken to develop a better understanding of the architecture and pottery of one of Morocco's most important archeological sites dating to the early Islamic period. The city of al-Basra was established by the Idrisids, the first Islamic dynasty in Morocco. This work is well underway with the excavation of stone structures, occupation floors, and brick kilns, and the recovery of abundant pottery, animal bone, and metal and glass objects. It is expected that future excavations at al-Basra will continue to yield rich cultural remains of one of Morocco's earliest Arab cities.

This monograph provides a description of the documentary and archeological evidence for al-Basra, including an account of the city's history, site stratigraphy, pottery classification, and pottery chronology. In addition, it offers a reconstruction of the pottery industry at al-Basra. The industry consisted primarily of small, independent artisan workshops that produced utilitarian pottery for urban inhabitants. The industry appears to have been technologically conservative, with few innovations introduced during its 300-year-history.

I am grateful to many colleagues and friends in the United States and in Morocco who provided encouragement, support, and facilities during the course of this research. Dr. Redman first offered me the opportunity of working in Morocco. Dr. Vincas Steponaitis provided encouragement and advise in the formulation of the data and theories presented in this work. Dr. Margaret Conkey offered lively discussions of archeological theory. Thomas Webster prepared most of the illustrations. Many other members of the faculty and graduate community at Binghamton provided intellectual stimulation and friendship over the years.

Mde. Joudia Hassar Benslimane, Chief of Archaeological Service in Morocco, graciously facilitated my research work in that country. Dr. Robert Shea, head of the American Legation Museum in Tangier, and other members of the American diplomatic community generously proffered a helping hand during my stay in 1983.

My research in Morocco was supported by grants from the National Science Foundation (BNS-83-11644), Sigma Xi Society, the Research Foundation of the State University of New York at Binghamton, and by various fellowships from the Department of Anthropology, State University of New York at Binghamton. The preparation of this manuscript was undertaken during my tenure as a Post-Doctoral Fellow in Materials Science at the Conservation Analytical Laboratory, Smithsonian Institution.

I am especially grateful to my parents, Fred and Anne Benco, my family, and many friends, particularly Ruth Smith Bolster, Mary and John Hibbits, Arlene and Harry Markowicz, Caroline Steele, and Laurie Steponaitis, who have given me their unswerving encouragement and support during the last few years.

# CHAPTER 1

# INTRODUCTION

## Objectives of the Research

This study has three major objectives. First, it confronts the broad theoretical issue of political and economic interaction in state and imperial societies by examining one aspect of the relationship, the nature and extent of political/administrative involvement in the production of craft goods. Toward this end, it utilizes documentary and archeological data to develop several possible explanations for changes in craft production at a medieval Islamic urban center.

Second, the study addresses the methodological problem of measuring and interpreting ceramic variability in the archeological record. It develops an approach to examine one aspect of that variability, that is, the diversity of pottery types in an archeological assemblage. The method distinguishes between functional and technological diversity in order to isolate the changes that are related specifically to production factors and, in a rough way, to the degree of competition in a local pottery industry.

Third, the study fills a gap in our current knowledge of medieval North African pottery by developing a typology and chronology for ceramic material from the first systematically excavated, multicomponent site dating to the early Islamic period in Morocco. The ceramic material from the site of al-Basra serves as the basis for a comparative study of pottery from other early Islamic sites in the western Mediterranean.

## Previous Approaches to the Study of Craft Production in Complex Societies

In the past two decades, archeological researchers have devoted a great deal of attention to the role of the economy in the development and maintenance of early state and imperial societies. At the beginning of this period, archeologists focused on the role of intensified agricultural production in the creation of centralized, administrative bureaucracies (e.g., Wittfogel 1957; Sanders 1968). In the 1970s, they shifted their attention to the role of long-distance and interregional trade in the formation of state societies (Flannery 1972; Johnson 1973; Rathje 1971; and Wright and Johnson 1975).

More recently, investigators have begun to consider the importance of non-agricultural or craft production in the evolution and maintenance of state and imperial societies. To a large degree, they have been concerned with defining the nature and extent of political/administrative involvement in the production of craft goods, using archeological material from the Near East (Johnson 1973; Wright and Johnson 1975), North Africa (Myers 1984), Mesoamerica (Brumfiel 1980; Feinman 1980, 1985; Feinman et al. 1984; Rice 1977, 1981; Spence 1981), and the Andes (Morris 1974, 1982; Morris and Thompson 1970). With a few exceptions (e.g., Morris, Myers, Spence), researchers have examined this issue from a systemic, processual perspective, with a view of artisan production as an integral and closely interconnected part of the larger political and economic system.

In many respects, the conceptual framework for these studies, which was inspired by Childe (1950), was set forth by Wright and Johnson in their research on the origins of the state in Mesopotamia (Johnson 1973; Wright and Johnson 1975). Using archeological data from southwestern Iran, they argued that the emergence of the state was associated with a sweeping reorganization of the local economy, which included both production and distribution systems. They contended that a growth in political centralization was accompanied by a shift in the organization of craft production from small, dispersed workshops to large, centralized shops located at administrative centers. At the same time, they found that technological processes changed from the manufacture of individually crafted items to mass-produced goods. Finally, they argued that the distribution of craft goods also shifted from local to regional exchange. Wright and Johnson speculated that these changes were instituted by the local ruling elite in order to meet the unpredictable fluctuations in demand for craft products, as well as agricultural goods, generated by the appearance of large groups of specialized nomads on the Susiana plain (Johnson 1973: 159; Wright and Johnson 1975:283). Although these hypotheses have yet to be tested with archeological data from Mesopotamia, Wright and Johnson's work outlined the basic premises of a politically administered economy for later archeological studies. These were: (1) that political factors may have an impact on craft production, (2) that changes in political organization may stimulate concomitant changes in the organization and technology of craft production, including pottery manufacture, (3) that a growth in political centralization often is associated with an increase in the scale and speed of production, and (4) that political/administrative involvement in craft manufacture can be detected in the archeological record by a study of the material objects themselves.

The ideas proposed by Wright and Johnson were subsequently elaborated upon by Feinman and applied to an archeological situation in the Oaxaca Valley of Mexico (Feinman 1980, 1985; Feinman et al. 1984). Drawing primarily on regional studies conducted in Imperial China (Skinner 1977a, 1977b) and modern-day Guatemala (Smith 1976), Feinman contended that an increase in political consolidation in the Oaxaca Valley, and thus stronger administrative control over the economy, led to a growth in the scale of craft production and a reduction in the amount of competition among producers. Conversely, an increase in regional political fragmentation, and thus weaker administrative involvement in the economy, resulted in a decline in the scale of craft production and an increase in competition among craftspeople. Feinman postulated that a growth in political control of craft production and distribution occurred during periods of valley-wide food shortages. Following along the lines of an argument made by Brumfiel for the Late Aztec period in the Valley of Mexico (Brumfiel 1980), Feinman suggested that Oaxacan administrators "monitored and manipulated" the production of craft goods in urban centers in order to encourage intensive, specialized agricultural production in rural areas (Feinman 1985:209).

Feinman tested this model on surface-collected ceramic material covering a 1,000-year period in the Oaxaca Valley, using several archeological indicators to measure changes in production scale and in degree of competition among producers. As the model predicted, he found that periods of regional political consolidation were associated with strong administrative control of pottery production, as reflected by a growth in production scale and a decrease in competition among potters; periods of regional political fragmentation were correlated with weak administrative involvement in ceramic production, which resulted in a return to small production units and a growth in competition among potters.

The basic premises set forth by Wright and Johnson also underlie Rice's work on the evolution of specialized pottery production in Mesoamerica (Rice 1977, 1981). Rice postulated that an increase in political centralization, i.e., the emergence of ranked and stratified societies, resulted in a concomitant increase in administrative involvement in pottery production. She argued that an increasingly powerful ruling elite provided the conditions that led to the development of specialized pottery production, in that they restricted access to valuable resources, such as special clay beds. Rice suggested that administrative control initially extended to the production of elite wares and later, when political power became more consolidated, it involved the manufacture of utilitarian pottery which the ruling elite extracted from artisans as tribute. In contrast to Feinman, Rice argued that as political centralization grew two trends in production emerged, according to the type of ware involved: (1) there was an increase in competition among potters making elite wares and (2) there was a decline in competition, as well as an increase in the scale of production, among potters making utilitarian wares.

## Limitations of the Feinman-Rice Models

There are several shortcomings in the models proposed by Feinman and Rice for reconstructing the organization and technology of ceramic production in complex societies. The first is the assumption that political organization and craft production, specifically pottery manufacture, are usually linked in a direct manner. As Chapter 2 will show, historical evidence from the Roman and Islamic periods in the Mediterranean indicates that the state was directly involved only in the production of certain crafts that were required for its own use. These manufactured products included textiles for the ruling elite, the royal palaces, and diplomatic gifts; armaments and, sometimes, clothing for the military; and money for exchange. Only occasionally was the state directly engaged in ceramic production and then only for specific purposes, such as supplying military troops or furnishing royal palaces. As a study by Morris shows, a similar situation prevailed at Inca administrative centers in the Andes, where the state manufactured pottery vessels specifically for use at state banquets and, possibly, for state beer (chicha) production (Morris 1974, 1982). This pottery, which is stylistically and compositionally uniform, has been found almost exclusively at administrative centers, like Huánaco Pampa, and rarely appears at sites in the surrounding countryside. In any case, there is little historical evidence for direct government involvement in the production of manufactured goods for consumption by the general populace in the Andes or elsewhere.

A second, related shortcoming is the assumption that a change in political organization necessarily leads to a shift in the organization and/or technology of pottery production. This assumption fails to consider two possibilities: that political organization may vary without a concomitant change in the structure of craft production and, more importantly, that the organization and technology of a craft industry may change primarily in response to shifts in consumer needs and demands, which in turn may stem from a number of demographic, cultural, or socioeconomic factors. For example, the arrival of immigrants who traditionally cook with flat ceramic griddles might lead to a demand for such products in an area where they had not been available previously. Likewise, an influx of rural peasants into an urban center might result in an increased demand for large quantities of cheaply made utilitarian wares. The actual mechanism for such changes is the introduction of new techniques by incoming craftspeople or the development of new methods by indigenous artisans. In any event, technical and organizational changes in a ceramic industry may occur entirely in response to economic factors, without the interference, direct or indirect, of the state or its agents.

A third drawback is the assumption that state-directed ceramic production usually leads to an increase in the scale of production and a reduction in the amount of competition among potters, as manifested by highly standardized, mass-produced pottery assemblages. Historical evidence from the Mediterranean suggests, on the contrary, that government-sponsored production may be undertaken by small production units, such as individual artisans who produce an entire product, but who work within the confines of a state workshop. In addition, such production may involve a high investment of time, in that the manufacture of elite wares by technologically advanced processes usually necessitates an increased number of production steps, such as in the case of lusterwares from Islamic North Africa (Jenkins 1980) or slipped wares from the Mayan site of Barton Ramie (Rice 1981). On the other hand, commercial production itself may be characterized by large-scale production units and a monopoly of one or two successful workshops in an area, such as the potteries at Arrezo in Roman Italy.

A final shortcoming with this approach is a general failure to consider the cultural context of pottery, that is, to take into account the function and value of ceramic goods within the particular society that made and used them. To a certain extent, Rice (1981) does this in her study of Mayan pottery by drawing a distinction between elite and utilitarian wares and, indeed, finds that they changed in different ways over a period of time. In a broader sense, however, the value and use of pottery depend on cultural attitudes toward the general category of ceramic vessels. These attitudes may be formed by such closely interrelated factors as the quality of local pottery-making resources, the technological capabilities of the artisans themselves, or the availability of other materials, such as metal, animal skins, or gourds, for use as containers.

## An Alternative Approach

An alternative approach to that proposed by Feinman and Rice is a free-enterprise, or entreprenurial, model. The free-enterprise scenario views specialized pottery production in state and imperial societies as consisting largely of commercial production, involving independent artisans who work in their own shops. This scenario attributes changes in the technology and organization of ceramic manufacture primarily to shifts in the marketplace, that is, in consumer needs and demands rather than to state interference in or control of pottery making. Such changes in the marketplace may stem from population movements, rising or falling socioeconomic levels, or environmental fluctuations. Finally, this approach recognizes that, in some cases, specialized ceramic production may come under state influence but that this consists of a relatively small portion of the total pottery production in a society and is usually limited to specific vessel classes, e.g., elite or military pottery.

This approach will be further developed and tested in the chapters that follow.

Chapter 2 presents the documentary data that describes the nature and organization of craft production in the Roman and Islamic periods in the Mediterranean. It includes a discussion of the different forms of specialized ceramic production present in the Islamic world.

Chapter 3 provides a historical overview of the development and decline of the early Islamic city of al-Basra in northern Morocco. The city constitutes the case study for this research. The account is drawn primarily from documentary and numismatic evidence and places al-Basra within its wider regional political and economic context.

Chapters 4 through 7 present the archeological evidence uncovered at the site of al-Basra during excavations conducted in 1980 and 1981 by the American-Moroccan Archeological Mission. Chapter 4 contains the stratigraphic and architectural evidence from al-Basra. Chapter 5 describes the cultural material, principally the pottery, found at the site. Chapter 6 concerns itself with depositional issues. Chapter 7 presents a ceramic chronology for al-Basra and compares the al-Basra pottery with ceramic material from other early Islamic sites in the western Mediterranean.

Chapter 8 describes the results of a diversity analysis of al-Basra pottery types in terms of the model of ceramic production presented in Chapters 2 and 3. The study is used to monitor changes in the scale of production and the amount of competition among potters at al-Basra and to identify some of the factors involved in its change through time.

Finally, Chapter 9 summarizes the findings of the study of pottery manufacture at al-Basra and concludes with a discussion about the nature and organization of craft production in state and imperial societies.

# CHAPTER 2

# HISTORICAL VIEW OF CRAFT PRODUCTION

This chapter examines the documentary evidence available for the organization of specialized craft production in early Islamic society. Although it is primarily concerned with pottery manufacture, it discusses other forms of craft production in order to present the broad spectrum of artisan activity in an urban society. The chapter focuses on three aspects of craft manufacture: (1) the relative value of various crafts within a culture, (2) the basic forms of labor organization, and (3) the political and economic contexts of craft manufacture. This study also draws upon comparable data from the Roman period in order to underscore the strong continuity in the basic features of artisan production over a period of 1,000 years in the Mediterranean area. It is important to note that, on the whole, craft production played a relatively minor role in the political economies of both societies (Finley 1973:139; Goitein 1967:116; Jones 1974:37).

## Islamic Craft Production

The most important source of documentary information about craft production in the early Islamic period in the Mediterranean is the Cairo Geniza. This remarkable collection of documents dating to the tenth to twelfth centuries includes official, business, and private correspondence written by merchants and artisans from all over the Mediterranean and deposited by them in a synagogue and cemetery in Fustat (Old Cairo), Egypt. Although they were written primarily by Jewish businessmen, the documents are generally believed to reflect the wide range of economic activities and arrangements that existed at the time. According to S. D. Goitein, whose five-volume series _A Mediterranean Society_ is based on these accounts, Jewish traders and craftsmen not only lived in close contact with Muslims and Christians, but they frequently formed partnerships and cooperated in other ways with them (Goitein 1970:51-52). In addition to the Geniza papers, other textual information on Islamic craft production comes from various travel books, histories, and literary works, some of which have been translated or compiled by others (e.g., Mez 1937; Marzouk 1955).

### Cultural Attitudes Towards Islamic Crafts

According to documentary sources, the most prominent industry in the Islamic world was the production of textiles. The highly specialized industry was so large, in fact, that it may have employed more than half of the working population, among them spinners, weavers, and dyers (Goitein 1967:101). Consisting primarily of silk, linen, and cotton, textiles were used for clothing and home furnishings, such as carpets, tapestries, and cushions. As the following observation indicates, textiles were highly valued in Islamic society:

> Of the three basic necessities of the human body, food, clothing and housing, clothing was the most important to the dwellers in the Near East. The clothier's art was more elaborated than

> any other, and interior decoration consisted almost entirely in coloured hangings for the rooms. Luxury meant to have handsome carpets on the walls and on the floors [Mez 1937:459].

The high value placed on clothing and on home furnishings was derived primarily from their beauty and durability, and thereby their investment potential.

> Clothing formed part--sometimes a considerable part--of a family's investment, being transmitted from parents to children, to be converted into cash in case of an emergency. The furniture of a house consisted mostly of various types of carpets, couches, cushions, canopies, and draperies--objects produced mainly by the textile industry. All this explains why the latter was of such paramount importance [Goitein 1967:101].

Next to textiles, the production of gold, silver, and copper objects was a prominent industry in the Islamic world. According to the Geniza papers, gold and silver ornaments, in addition to being decorative, also served as a form of investment and savings (Goitein 1967: 108). They were often used as payment in commercial transactions and were prominent items in dowry lists, particularly among Jews. Although copper items were less valued, they occasionally appeared in trousseau lists, generally in the form of better kitchen utensils. Other important craft industries included the manufacture of glass, leather, particularly for footwear and shipping containers, and paper products.

In contrast, pottery production is mentioned relatively infrequently in the Geniza papers. It appears to have been a fairly specialized industry, from the variety of names given to potters. *Fakhkhari*, for example, were potters who made clay pipes and other clay materials used for construction. *Qaddari* were potters who made vessels for serving or storing food and drink. The latter could further specialize in the production of a particular type of vessel, e.g., *kuzi* made narrow-necked water jugs without a spout, and *ghada'iri* made porcelain-like fine earthenware dishes (Goitein 1967:110-111).

Furthermore, the value of pottery in the Islamic world varied, depending on its origin and method of manufacture. Imported Chinese porcelain was highly prized, as attested to by its inclusion in wealthy trousseaus and its use as a presentation gift (Goitein 1983: 146). Fine glazed earthenwares (e.g., *ghadar*), which served as popular imitations of porcelain, were also in great demand throughout the Mediterranean. As a number of documents testify, fine earthenwares were shipped long distances both by sea and overland by camel, for example, from the Maghreb to Egypt and from Egypt to the Persian Gulf (Goitein 1983:146). On the other hand, unglazed utilitarian wares, such as the ubiquitous clay storage jars, pitchers, bowls, and lamps found at archeological sites throughout the Maghreb, were not mentioned at all in the Geniza papers. They were not included in dowry lists nor listed in commercial documents, which suggests that they were considered of relatively little value and were probably not traded any distance. On the whole, unglazed pottery seems to have been held in relatively low esteem during the early medieval period in the western Mediterranean.

## Forms of Craft Organization

The documentary evidence identifies four basic forms of labor organization in specialized craft production.

By far, the most common form was a small, independent workshop, which was owned and operated by an individual artisan, a family, or a small number of partners, usually numbering no more than five (Goitein 1967:81). Although they varied, partnerships usually involved an arrangement in which all partners shared labor, capital investments and profits equally, or in which some partners contributed capital and others provided labor (ibid:170). A small workshop rarely, if ever, employed wage labor. Goods produced in a small workshop were generally sold directly to consumers in the shop or through wholesalers in local markets. During the Geniza period, a variety of artisans operated small workshops, among them goldsmiths, silversmiths, coppermakers, glassmakers, silk weavers, and pottery makers.

A variant of this form was a nucleated industry, in which a number of individually operated workshops were grouped together to form a more or less tightly clustered artisan complex. Although it is not mentioned in the Geniza documents, this form is evident at some early Islamic sites, such as Siraf in the Persian Gulf (Whitehouse 1970). At Siraf, a number of small kilns and potting sheds were found clustered together in an enclosed compound in the city near the edge of the sea; its location suggests that pottery may have been loaded directly onto ships and transported to distant markets. While encouraging cooperation among craftsmen, a nucleated industry also tended to attract middlemen who operated relatively wide distribution networks. As we will see later, nucleated potteries were prevalent during the Roman era (Peacock 1982).

A less common form was a large commercial workshop or factory, which was owned by a group of partners, most of whom were primarily investors (Goitein 1967:81). Large independent workshops often employed wage labor, although the number of workers and the conditions of their employment are not mentioned in the Geniza papers. During the early Islamic period, large workshops seem to have been limited primarily to sugar and paper production, both industries which involved technologies that were imported into the Islamic world from, respectively, India and China. Immediately after its introduction in the mid-eighth century, paper was produced in factories owned by the elite, e.g., governors and sultans, and only later was its production taken over by middle-class entrepreneurs (ibid:81). By contrast, sugar manufacture was mostly in the hands of private producers until it became a government monopoly under the Mamluks in the thirteenth-century (Ashtor 1977). Partly because they involved technologies with no local tradition, these sugar and paper factories employed relatively large numbers of unskilled workers (Goitein 1967:81).

A fourth form was a large, state-operated workshop or factory which produced goods for government consumption. During the Geniza period, state industries concentrated on the manufacture of textiles, armaments, and currency (ibid:81-82). State workshops usually employed skilled craftsmen who worked relatively independently and were able to form partnerships among themselves. In some cases, skilled artisans were forced to work in state-run operations. In a letter dated about 1040 A.D., an artisan in Damascus complained that the local government recruiter had chosen the best craftsmen for the state textile factory and, once in, they could not get out except by petitioning the Caliph, which was viewed as a rather extreme measure (ibid:82). Although most government production took place in centralized workshops, some operations were dispersed among workers' homes, similar to a contemporary cottage industry. In Egypt, for example, women spun flax in their homes, under the direction of government agents who provided materials, set standards, and purchased the finished product (Mez 1937:461). Two early documents provide a glimpse of the government's role in two aspects of textile production--the processing of raw materials and the weaving of the cloth. The first is a description by the twelfth-century traveler Ibn Balkhi:

> The flax is soaked in ponds, then pulled apart and spun into threads. The linen thread is washed in the water of the Rahban Canal; though the water of this canal is scanty, it has the property of bleaching the linen thread, which becomes white in no other water. This Rahban Canal belongs to the royal treasury, and its earnings now belong to the house of the Emir, since the exchequer grants the use of it only to those weavers who are commissioned by it to weave. A fiscal inspects as overseer, and brokers fix the proper price of the fabrics, sealing the bales before they are put into the hands of foreign traders. These relied on the brokers, and purchased the corded bales, just as they were lying, and in each town whither they were brought, people merely asked for the certificate of the broker of Kazrun, and sold the bales without undoing them. Hence it often came to pass that a load of bales from Kazrun changed hands more than ten times without being opened [Mez 1937:462].

The second is a fourteenth-century account of a Sultan's visit to a textile factory in Alexandria, Egypt:

> He [the Sultan] was conducted to Dar at-Tiraz (the textile factory) by his minister. He dismounted, entered and went up the staircase to the place where there were the looms and the stores. He observed every weaver at his loom weaving various decorated textiles and complete clothes (_badlat_) of different colours, already folded for the use of the harem of the Sultan. He went around the looms, observing them and putting his head below them to see their lower parts and enjoyed himself by looking at the weavers while they were working and throwing their shuttles to and fro. . . The Sultan then saw what was in the Dar at-Tiraz of Alexandria namely: the _zarakish_, _raqm_, and golden silk robes completely finished. He chose some of these to be sent with him and left the rest until they should be completed [Marzouk 1955:65-66].

## Political/Administrative Context of Craft Production

As the discussion above suggests, Islamic government involvement in craft production was limited primarily to industries that were considered vital to the operation of the state, namely goods required by the imperial court, government administrators, and the military. State run textile factories made clothing and other goods, such as banners, tents, and palace furnishings needed by the rulers and their families; in addition, they produced special cloths and robes of honor for presentation as diplomatic gifts. Government mints produced currency from various gold and silver bars, vessels, and old coins and other fragments. Instead of issuing money per se, however, the mints obtained these metal items from suppliers (murid) and returned the equivalent to them in currency, less the cost of labor and a government fee (Goitein 1967:267).

The extent of the government's involvement in private craft production appears to have been minimal. In general, it was directed toward raising revenues. The government collected taxes from private workshops in the form of a monthly license fee and it imposed duties on goods entering or leaving a city (Goitein 1967:269-271). The government also appointed market supervisors ('arif) to guard against fraudulent practices in the marketplace, such as overcharging or selling defective goods. The government occasionally took possession of commodities that were being shipped from one place to another, although it usually reimbursed the merchants for them. According to the Geniza papers, the government confiscated such goods as fine textiles, lead, wax, and olive oil, all items which the imperial court or bureaucracy presumably required. In summarizing the attitude of merchants toward state seizures of this sort, Goitein wrote:

> The tenor of the letter reporting this occurrence clearly shows, however, that the writer would have welcomed selling his merchandise to the government, and the relatively rare occurrence of references to the forceful seizure of goods by the government agencies seems to indicate that these practices were not particularly oppressive [Goitein 1967:268].

The state's involvement specifically in pottery production is not well documented. There is, however, some evidence to suggest that the government engaged in ceramic manufacture from time to time, albeit indirectly, in order to furnish imperial palaces and other government buildings and possibly to supply stationary military forces. It appears that the state obtained pottery for these purposes by commissioning artisans or by requisitioning or purchasing them from private workshops rather than by operating its own pottery workshops. On the whole, the government's involvement seems to have been intermittent and opportunistic and did not represent an ongoing commitment or policy.

There is some reason to believe that the ruling elite occasionally commissioned or purchased ceramic material to decorate or furnish royal palaces and other government buildings. This material consisted primarily of luxury wares, i.e., pottery that was highly decorated and/or made by an advanced technological process. Textual and archeological evidence from the ninth century indicate that the governor of Ifriqiya (present-day Tunisia) recruited a Baghdad potter to manufacture luster-painted tiles for the Great Mosque at Kayrawan, thereby introducing the lusterware technique into the Maghreb (Jenkins 1980; Marcais 1928). Similarly, an eleventh-century Zirid ruler brought potters, possibly from Bougie in Algeria, to make luster-painted tiles for the new royal palace in Malaga (Jenkins 1980). The tenth-century Umayyad Caliph in Cordova recruited or commissioned hundreds of craftspeople from the Middle East to decorate the newly built palaces at Medina az-Zahra (Dozy 1972:446); among them apparently were potters who introduced or developed several new techniques, including _cuerda seca_, which appears in Spain in the earliest levels of Medina az-Zahra (Zozoya 1980b:314).

There is less evidence for state-sponsored production of military pottery in the early Islamic period, although an argument has been made for military involvement in pottery manufacture at Qsar es-Seghir in northern Morocco during the fourteenth and fifteenth centuries (Myers 1984). Considering the nature of Islamic armies in the Maghreb during the centuries immediately following the conquest, it seems unlikely that the military would have been engaged in or even interested in making pottery. In contrast to large, standing Roman armies posted in forts along the imperial frontiers, Arab armies, particularly in the Maghreb, were composed of highly mobile Berber troops. Because of their mobility, it seems likely that the armies would have used durable, lightweight equipment, such as metal cooking and eating utensils, rather than fragile ceramic vessels, during their campaigns. In their garrisons, which were often located in urban centers, they would likely have had access to commercially produced pottery for domestic purposes.

## Roman Craft Production

A number of striking parallels can be seen in the organization of craft production between the Islamic period and the earlier Roman period. The basic modes of craft production were similar, although some forms appear to have been more prevalent in the Roman period. The political context of production, i.e., government involvement in textile and armament factories and mints, also seems to have been very much the same in the Roman world.

In addition to abundant documentary information (Haywood 1956; Waltzing 1968), the evidence for craft production in the Roman world includes detailed archeological data, in particular the well-excavated remains of ceramic workshops, kilns, and pottery molds and stamps (Peacock 1982; Hayes 1972). Together, documentary and archeological data have been brought together in several economic histories, which form the basis of the following discussion (e.g., Finley 1973; Hopkins 1978; Jones 1974).

### Forms of Roman Craft Organization

Like that in the Islamic period, the most common form of craft organization in the Roman world was a small, independent workshop operated by one or a handful of individuals. These workshops were situated in most towns and cities where they served the needs of local inhabitants, including landowners, farmers, and other craftspeople (Jones 1974:38).

Very prominent in the Roman period was the nucleated industry, in which a number of small workshops and/or moderately large-scale commercial workshops were clustered together in an urban or rural area. Their location depended to a large extent on proximity to sources of raw material, such as wool or clay beds, and to transportation networks. Nucleated workshops usually specialized in the manufacture of technologically superior, medium-quality products, such as fine red-slip tablewares (known variously as Arretine, samian, and terra sigillata), or finely woven textiles (e.g., Tarsus linen). Because they were produced on a large scale at a relatively low cost, these goods could be distributed more widely than the products of a single workshop. In fact, as Peacock points out, craft centers attracted middlemen with their wide distribution networks (Peacock 1982:127). Some of the best evidence for nucleated industries comes from archeological excavations of potteries in both the imperial center and provinces of the Roman empire. The nucleated pottery industry in Arezzo in northern Italy, for example, consisted of small- and moderately large-scale workshops that seem to have operated fairly independently of each other, as the layout of the facilities suggests (ibid:121). Some of the large workshops often employed slave-artisans, as the names found on pottery stamps indicate. The largest establishment, owned by a man named Rasinus, employed up to 60 slaves, although they probably did not all work there at the same time (ibid:122). Most larger workshops seem to have operated with 10 or fewer workers. The nucleated industries in central and southern Gaul, on the other hand, consisted almost entirely of small enterprises that appear to have operated without slave or wage labor, possibly because of a Celtic tradition of small workshop organization (ibid:122-123). Many nucleated industries in Gaul were located near waterways that could provide cheap transport, such as those at Lyon and Lezoux. Others were situated in isolated rural areas, such as La Graufesenque, apparently to take advantage of the superior clay in the area. In Britain, nucleated potteries were often located on roads between major towns, such as those at Oxford.

Finally, another form of craft organization was the state-run workshop. Like that in the Islamic period, government workshops in the Roman period produced textiles, armaments, and coins--all items necessary for the operation of the imperial court, the bureaucracy, and the military. According to Jones (1974:403), state armament factories were manned by workers who were legally soldiers; government mints were largely staffed with imperial slaves. In the late Roman period (fourth and fifth centuries), workers in both industries became part of a hereditary caste system, in which individuals were free and could own property. Although it normally acquired military garments from commercial textile manufacturers through compulsory purchases or requisition, the Roman government began to operate its own textile factories for military clothing sometime after the fourth century when the army was increased in order to combat growing incursions on imperial frontiers (ibid:356). State workers in imperial woolen (gynaica) and linen (linyphia) factories were legally slaves, although by the late fourth and fifth centuries they too formed a caste system. In the imperial factories, the government provided workers with rations and raw materials and, in turn, each worker was required to produce a fixed quota of finished articles, e.g., military tunics or helmets (ibid:356).

## Political/Administrative Context of Roman Craft Production

As we saw above, the Roman government was directly involved primarily in the production of military equipment, imperial textiles, and coins. On the whole, its involvement in private craft production was limited to the collection of business taxes (*collatio lustralis*) and to the imposition of duties on goods in transit, very much like that in Islamic times. From time to time, the Roman government also instigated compulsory purchases or requisitioned goods that it needed, particularly military clothing and equipment (Jones 1974:356). There is also documentary evidence that the government periodically used craftsmen and other urban workers to provide corvée labor for local or state projects, but this work usually was not connected to the workers' trade (Jones 1974:50). Corvée labor, however, was often organized through the local associations of artisans and merchants (*collegia*), whose primary functions at that time were religious, social, and benevolent. Similar associations existed in the early Islamic period and only in the later Islamic period did they take on the regulatory and protective functions common to medieval guilds (Goitein 1967:84-85).

The Roman government seems to have been minimally involved in the manufacture of pottery. According to archeological evidence, the Roman army in Britain obtained pottery in several ways (Darling 1977). The army imported highly specialized vessels, like drinking cups, from the Continent. Individual soldiers or troop units purchased their own kitchen pottery, such as cooking vessels and mortaria, from local potters. In exceptional circumstances when troops could not obtain the quantity and quality of goods they needed locally, perhaps because they were located in an isolated area, the army provided skilled craftsmen. Most Roman military pottery in Britain, which varies considerably from one fort to another, as well as within individual forts, seems to have been made and acquired locally, although it is not completely clear whether the pottery was acquired by purchase or requisition or as part of a taxation process (Young 1977). One exception to this trend, however, may be the manufacture of black-burnished ware, which was produced in southeastern England but has been found archeologically at Roman forts in Scotland, far to the north. According to Williams (1977), the distorted distribution of this particular ware may indicate that the pottery was made and distributed under administered conditions.

## Historical Model of Islamic Craft Production

Drawing on the historical evidence presented above, it is possible to develop an alternative model of specialized craft production that is more closely applicable to an early Islamic context in North Africa. In addition, the model constitutes an alternative to the Feinman-Rice approach for reconstructing craft production in other early state and imperial societies. The underlying premise of the model is that most craft production in preindustrial societies is entreprenurial, that is, that it is more directly affected by market forces than by political factors, as the Feinman-Rice model contends.

One of the major tenets of the free-enterprise model is that a large part of specialized craft production takes place in small, independent workshops, which are operated by a few skilled artisans, often members of the same family. Most craft items produced in small workshops are sold directly to customers in the shop or in nearby urban or rural markets. Because they produce primarily for the needs of local consumers, small workshops are particularly sensitive to changes in the local economy or population. Thus, in response to a shift in the size, socioeconomic status, or ethnicity of the local populace, the number of small workshops may increase or decrease, a new manufacturing technique may be introduced, or the type of products may change to satisfy new customers. On the whole, however, the basic unit of production, that is, the small, independently operated workshop in which there is relatively little division of labor, is likely to persist unchanged through time.

A smaller amount of craft production occurs in large-scale workshops or in nucleated industries, which may consist of large and/or small workshops clustered together. These workshops are often situated near sources of raw materials, such as good clay beds. In addition, they are located close to large, urban markets or to efficient forms of transportation, such as waterways or roadways. The large quantity and moderately low quality of items produced in these workshops generally offset the additional cost of transporting them long distances. Usually, these goods are sold to middlemen, who then distribute the products to customers in distant towns and cities. Because their production is often geared toward faraway markets, these large-scale industries are particularly vulnerable to regional political or economic upheavals, which may disrupt or dislocate transportation routes. In addition, they rely heavily on middlemen to buy their goods at reasonable prices and to find markets for their distribution. Thus, in response to shifts in access to certain markets or in the reliability of middlemen, these industries may retool themselves for new markets, they may relocate entirely, or they may disintegrate into small, scattered production units. In large pottery manufacturing centers in contemporary Spain, for example, potters who were taken advantage of by middlemen left the center and resettled in small towns where they could sell their products directly to local customers (Vossen 1984:346). The dispersal of artisans from nucleated or large scale industrial operations also serves to spread technical knowledge throughout a region. Unlike small workshop units that are fairly resistant to change, large-scale workshops or nucleated industries, both urban and rural, are far more vulnerable and tend to break down into small units or disappear entirely.

Finally, a minor part of the total pottery production in an urban or state society takes place in state workshops or under some sort of government control. The pottery produced in this way consists almost exclusively of luxury wares for the ruling elite or of utilitarian wares for military troops. It is made by artisans working in government facilities or in private workshops under commission to the state. In most cases, state-sponsored pottery production is intermittent and does not constitute an ongoing government operation, as the production of luxury textiles and military armaments do. Changes in the nature and extent of state-sponsored pottery production occur mainly in response to specific political or military events, such as the construction and furnishing of royal palaces or the quartering of troops in areas where there are insufficient pottery supplies. On the whole, state involvement in commercial ceramic production is minimal and is limited to the collection of certain taxes and duties.

# CHAPTER 3

# POLITICAL AND ECONOMIC HISTORY OF AL-BASRA

The Islamic town of al-Basra in northern Morocco was founded, flourished, and fell into decline during a relatively short period of time (A.D. 800-1100). The Arab city was established in the early ninth century by the Idrisids, one of Morocco's earliest dynastic families. It prospered as an agricultural, administrative, and commercial center under the Idrisids and later, in the tenth century, under the Fatimid empire. Toward the end of the tenth century, the city was partly destroyed. Rebuilt, it served as a small agricultural center under the Zenata confederacy in the eleventh century. The city was probably abandoned sometime in the late eleventh or the twelfth century during the period of Almoravid hegemony in Morocco.

This chapter traces the history of al-Basra within the context of three major political developments in the Maghreb during the early Islamic period (eighth through twelfth centuries): the formation of the Idrisid state; the expansion of the Fatimid empire into the Maghreb and its subsequent conflict with the Umayyads of Spain for control of northern Morocco; and, finally, the political fragmentation of the Maghreb on the eve of the Almoravid invasion. Many of these events had a direct bearing on the course of al-Basra's history and seem to have affected the degree of political centralization that prevailed at the urban center and in its hinterland during this period. The available evidence suggests that the amount of administrative influence at al-Basra was strongest during the late Idrisid and Fatimid periods, when the city served as a provincial capital and the impressive city walls were built.

## Documentary and Numismatic Evidence

The history of al-Basra is derived from a number of Arab chronicles written between the ninth and fourteenth centuries (al-Bakri 1965; Ibn Abi Zar 1860; Ibn Hawkal 1842; Ibn Idhari 1901-1904; Ibn Khaldun 1927; al-Idrisi 1969; Leo Africanus 1956; al-Mukaddasi 1950) and from a handful of coins that were minted at al-Basra itself (Brethes 1939; Eustache 1970-71; Lavoix 1891).

The early Arab histories are an important source of information about al-Basra. The histories consist of "traditions," accounts of events which have been transmitted from one person to another over a long period of time. Because of this method of transmission, they are subject to a certain amount of exaggeration, elaboration, and error. In many cases, the chroniclers present several "traditions" side by side, without providing any critical evaluation. For example, the eleventh-century chronicler, al-Bakri, offers several versions of the murder of Idris I by an Abbasid agent. In one account, Idris I is poisoned with civet and in another with a slice of melon (al-Bakri 1965:235 ff.). Al-Bakri makes no attempt to resolve the discrepancy. Such elaboration of detail is found more often in later "traditions" rather than in those written closer to the actual events.

Other important sources of information about al-Basra are Arab geographies, which provide descriptions of towns and cities in the Maghreb. Based on first-hand accounts or on the reports of other travelers, the geographies often reflect the personal biases of their authors. The descriptions of Ibn Hawkal (1842), who traveled through the Maghreb possibly as a Fatimid sympathizer, for example, reveal an unusual interest in the strength of city walls and the location of city water supplies. Despite their biases and inaccuracies of detail, however, the chronicles remain one of the most important sources of information on the political and economic climate of the Maghreb during the early Islamic period.

Further information about the city is available from six extant Idrisid coins struck at al-Basra in the late eighth and ninth centuries. Because they contain the dates and names of rulers at al-Basra at the time they were minted, they provide specific information about certain political events in the city. Unfortunately, the proveniences of these coins, which belong in private coin collections, are unknown. Two other coins were found at the site of al-Basra itself during the 1980-81 excavations but have not been dated because of their badly worn condition (see Figure 5.19).

Although fragmentary and sometimes contradictory in nature, the documentary and numismatic evidence forms the basis of a short history of al-Basra compiled by Daniel Eustache, a Moroccan historian and numismatist (Eustache 1955, 1970-71). The following account is drawn in part from Eustache's work and in part from an unpublished history of Morocco written by Schulman (n.d.).

## Formation of the Idrisid State in Northern Morocco

Located between the Atlantic coast and the Rif Mountains in northern Morocco (Figure 3.1), the Islamic town of al-Basra was established during a dynamic period of imperial fragmentation and state formation in the western Mediterranean. Less than half a century after their conquest of North Africa and Spain, the Arabs began to lose control of the western periphery of their empire--for several reasons. After overthrowing the Umayyad Caliphate in 750, the ruling Shi'ite Abbasids moved the imperial capital from Damacus to Baghdad, thereby shifting the empire's political and economic center of gravity from the Mediterranean to the Persian Gulf and Indian Ocean. In addition, the Arabs failed to fully subjugate the fiercely independent Berber tribes in North Africa, which continued their rebellion against imperial authority.

Within this political vacuum, a number of independent kingdoms emerged in the Maghreb. They were established by political and religious dissidents who fled to the Maghreb to escape Abbasid persecution (Figure 3.2). In 756, the Umayyad prince, Abd al-Rahman I, established the Merwanid kingdom in southern Spain (al-Andalus), with its capital at Cordova; the kingdom became a center of Sunni religious orthodoxy. In 757, followers of the Kharijites founded the Maghrawa dynasty at Sigilmasa in southern Morocco. In 777, the Kharijites also established the Rustamid kingdom at Tahert in central Algeria.[1] In 800, Ibrahim

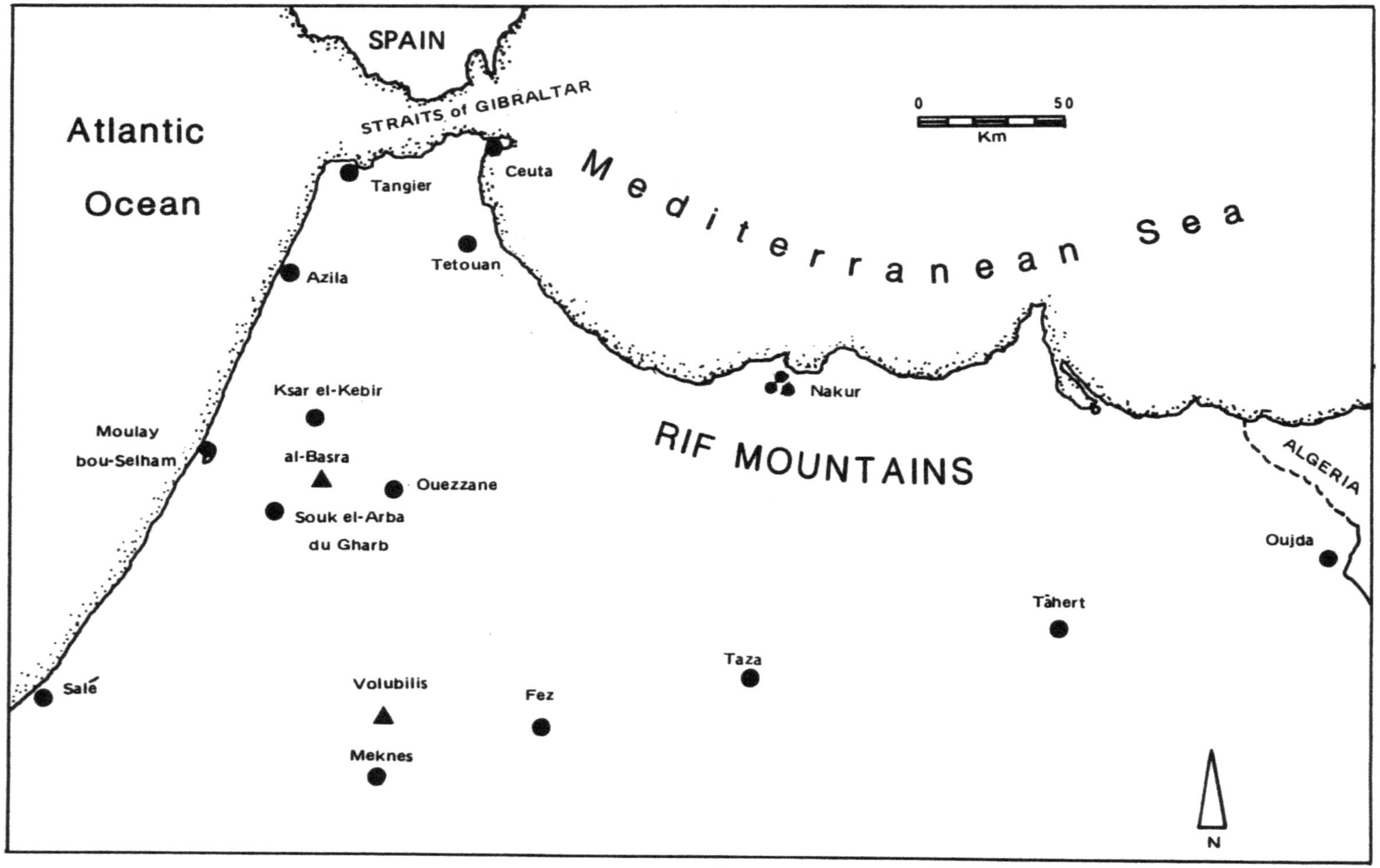


Figure 3.1.   Map of northern Morocco.

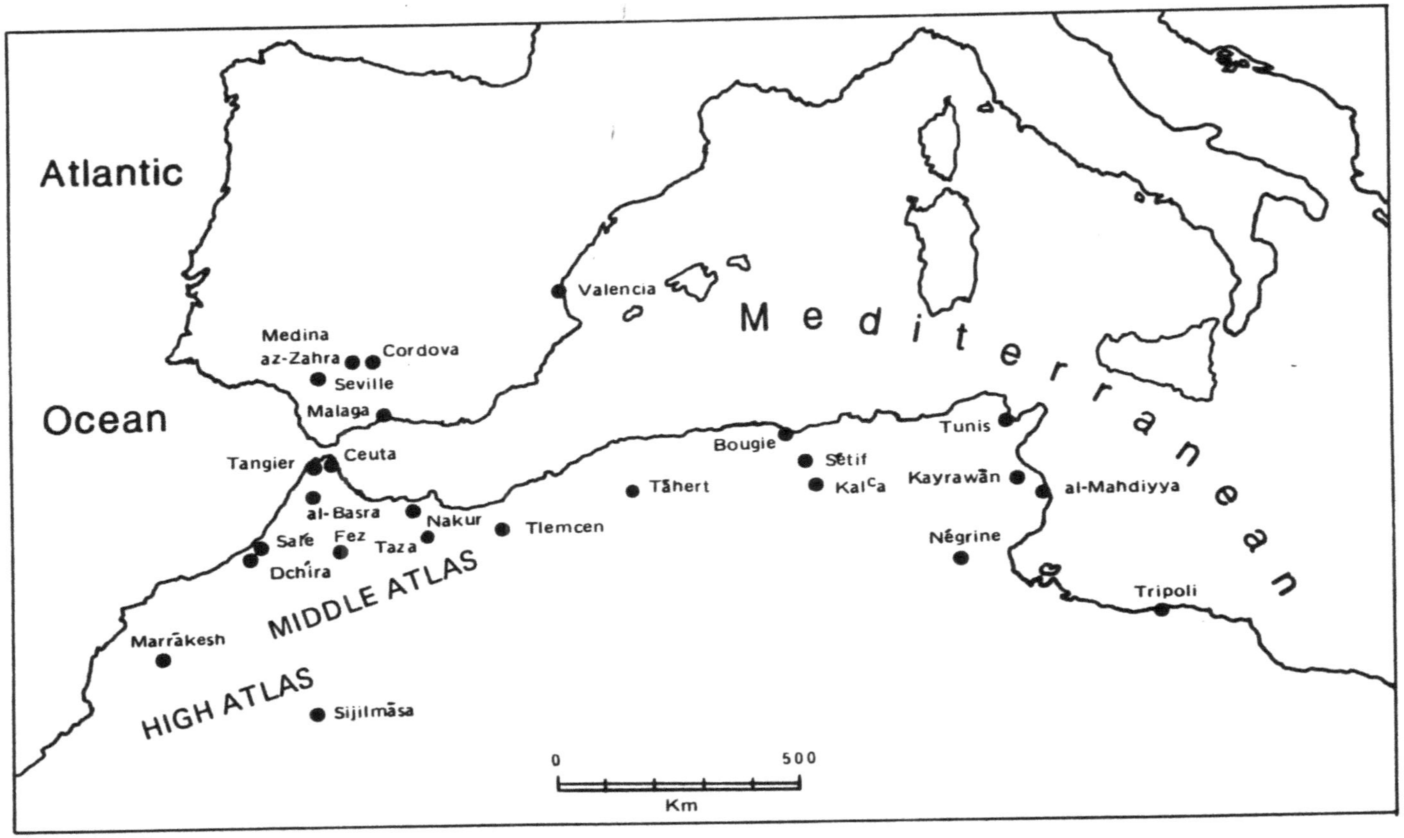


Figure 3.2. Map of early Islamic sites in the western Mediterranean.

ibn al-Aghlab, the Abbasid-appointed governor of Ifrikiya (Tunisia), set up the Aghlabid dynasty by naming his son to succeed him while he continued to pay tribute to Baghdad. In 789, the Alids, descendents of Ali, founded the Idrisid kingdom in northern Morocco, with its capital at Fez; it was this dynasty to which the fate of al-Basra was most closely tied.

According to the chronicles, the founder of the Idrisid dynasty, Idris I, fled to the Maghreb after the Abbasids defeated members of the Alid family at the bloody battle of Fahkh near Mecca in 787. With his freedman Rashid, Idris I settled in Walili, the ancient Roman town of Volubilis, in northern Morocco. In 789, he was proclaimed leader of a confederacy of local Berber tribes (the Awraba), laying the foundation of an independent state in northern Morocco. It has been suggested that his ability to unite the Berbers may have stemmed from his direct descent from the Prophet (<u>sharif</u>) and charismatic power (<u>baraka</u>), although the beliefs of the early Berbers are not well known (Norman Stillman, personal communication). Idris I ruled for only four years before his death in 793 at the hands of an agent of the Abbasid Caliph in Baghdad. Although Idris I left no heir, one of his Berber concubines, Kenza, gave birth to a son a few months later. The boy was raised by Rashid and, at about the age of 11, he was proclaimed leader of the Berber confederation.

It was under Idris II's reign (ca. 804-828) that the Moroccan kingdom was expanded to its widest extent, stretching from the Straits of Gibraltar on the north, the Atlantic coast on the west, the Sus river on the south, and central Algeria on the east. Idris II established a number of new cities, among them the royal capital at Fez and the town of al-Basra,[2] named after the ancient port in Iraq. Although one account claims that al-Basra was built on the foundations of a Roman city (Tissot 1877), this has not been borne out by archeological evidence. It is more likely that the city was established on the site of an agricultural village inhabited by Berbers (Eustache 1970-71:119). Idris II also Arabized the region, bringing Arab supporters from al-Andalus and Ifrikiya to settle in the new kingdom.

The centralized state established by Idris II dissolved at his death in 828. His eldest son, Mohammed, divided the kingdom among his brothers, apparently on the advice of his Berber grandmother Kenza. Mohammed kept Fez as his royal residence. He gave the northern prov-

---------------------

[1] The Kharijite movement was the first religious schism in Islam. The Kharijites' principles of ascetism, austerity, and egalitarianism were particularly attractive to the Berbers in North Africa where they gained many converts.

[2] The dates of two coins struck at al-Basra indicate that the city existed during the time of Idris II. One coin was minted in 796 under Idris II's regency and the other was struck in 812 under his reign (Eustache 1970-71:184-185, Plate I).

ince of Tangier, Asila, and al-Basra to al-Qasim;[3] areas of the Rif mountains occupied by the Sanhaja and Ghomara tribes to Umar, the regions of Salé and Wazzekur in the south to Isa, and other provinces to Dawud and Yahya. Although the land partition was meant to prevent rivalry among the brothers, civil war broke out almost immediately. One of the brothers, Isa, repudiated Mohammed's authority. In response, Mohammed asked al-Qasim to punish the rebel. Al-Qasim refused, but another brother, Umar, agreed. After defeating Isa, Umar also turned on al-Qasim and confiscated his territory. Al-Qasim retreated to Asila and the city of al-Basra passed into the hands of Umar's family for a couple of decades.

Although politically divided, the Idrisid state remained relatively peaceful during the reigns of Mohammed's sons, Ali (836-849) and Yahya (849-863), in Fez. In the late 850s, Yahya undertook a major building program at the royal capital of Fez, which included the construction of two new mosques. At about the same time, Yahya restored the northern province, including the cities of Asila, al-Basra, and Kurt, to al-Qasim.[4] In 863, al-Qasim entrusted the government of al-Basra to his eldest son, Ibrahim. Following his father's death, Ibrahim took over the government of the northern province, which he ruled until 900, very likely from his residence at al-Basra.[5] It may have been during Ibrahim's long and peaceful reign that the impressive city walls at al-Basra (Figure 3.3) were built (Eustache 1970-71:120). Other parts of the city may have been reconstructed or expanded at the same time, as archeological evidence suggests (see Chapter 4). After Ibrahim died in 900, the northern province continued to be ruled by his sons until 917 when the Fatimids invaded the Maghreb.

---------------------

[3] Somewhat different divisions are given by fourteenth-century historians. Ibn Abi Zar (1860), for example, attributes al-Basra, Asila, al-'Ara'is, and their dependencies up to the Ourregha River to Yahya ibn Idris II, and Tangier, Ceuta, Hajar en-Nasr, Tetouan, and their dependencies to al-Qasim ibn Idris II. According to Ibn Khaldun (1927), Yahya received Asila, al-'Ara'is and their dependencies up to the border of the Ourregha. It is likely that some of this information is incorrect. Hajar en-Nasr, for example, was founded only in the tenth century.

[4] Three Idrisid coins struck at al-Basra indicate that the city and its territory passed from Umar's family back to al-Qasim's family sometime in the 840s or 850s. Two coins were struck at al-Basra under the name of Ali ibn Umar; one of them was dated 846 A.D. and the other's date is unclear. A third coin with the date of 859 A.D. was minted at al-Basra under the name of al-Qasim ibn Idris II (Eustache 1970-71:184-185, Plate I).

[5] An Idrisid coin was struck at al-Basra under the name of Ibrahim ibn al-Qasim in 883. This is the latest Idrisid coin known to have been struck at al-Basra (Eustache 1970-71:185, Plate II).

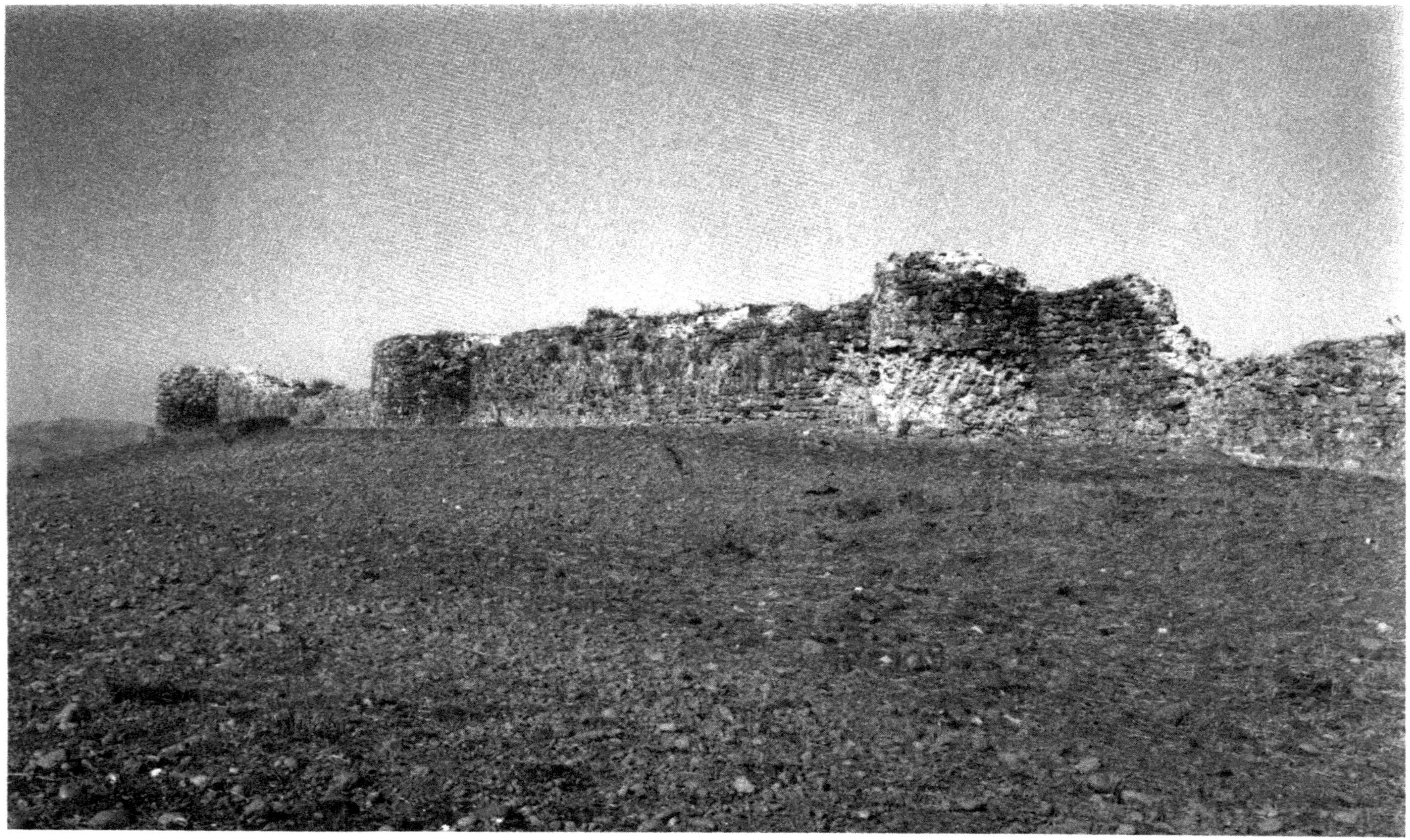

Figure 3.3. View of ancient city walls at al-Basra, looking east.

## Fatimid and Umayyad Imperialism in the Maghreb

During the tenth century, the Maghreb became a pawn in the struggle between the powerful Fatimids of Tunisia and the Merwanid Umayyads of Spain for control of the Far West. After overthrowing the Aghlabid dynasty in Tunisia (Ifrikiya) in 909, the Shi'ite Fatimids expanded westwards, establishing a new empire in North Africa. Between 909 and 921, they gained hegemony over the entire Maghreb, including the Rustamid, Idrisid, and Maghrawa kingdoms. According to some historians (e.g., Lacoste 1974), the Fatimid conquest was directed toward control of the long-distance trade in gold and slaves that had begun to flow through the Maghreb in the late ninth century. The cities of Fez, Sigilmasa, Tlemcen, Tahert, and Kayrawan had become major hubs in the trade network between West Africa and the rest of the Mediterranean. In response to the Fatimid threat in the Maghreb, the Umayyads of Spain seized the ports of Ceuta, Tangier, and Melilla during the next few years in order to ensure access to the sub-Saharan trade (Lacoste 1974: 3). The battles between these major powers in the Maghreb were generally fought by their Berber allies; the Sanhaja supported the Fatimids and the Zenatas aided the Spanish Umayyads.

Although the Idrisids became vassals of the Fatimids in northern Morocco, the first Fatimid governor Musa ibn Abu'l Afiya expelled them from their cities and chased them to Hajar en-Nasr, their refuge in the Rif mountains. When Musa defected to the Spanish Umayyads in 930, the Idrisids joined the Fatimids in ousting him from the region. Sometime between 934 and 938, the heirs of al-Qasim returned to their cities in the northern province, which they continued to govern under Fatimid authority. In 958, after the Idrisid prince Abu'l Ayyash in Asila defected to the Umayyads, the Fatimids sent their general Jawhar with an army of 20,000 Sanhaja and Kutama horsemen to reestablish Fatimid control in northern Morocco. According to one account, the Fatimids appointed a governor at al-Basra and made it the capital of the northern province for the next 20 years (Ibn Idhari 1901-1904). According to another account, Abu'l Ayyash's brother, al-Hasan, the last Idrisid prince, resided at al-Basra until 974 when he was expelled by the Spanish Umayyads and forced to retreat--with his harem and treasury--to Hajar en-Nasr (Ibn Abi Zar 1860). Sometime around 979, the city of al-Basra was destroyed--or its walls dismantled--possibly by the Fatimid vassal Buluggin ibn Ziri, who may have destroyed it in his attempt to restore Fatimid power in the Maghreb. This was the last campaign sponsored by the Fatimids. By 974, they had conquered Egypt, moved their capital from Kayrawan to the newly built city of Cairo, and had lost their interest in the Maghreb. The specific cause and the extent of destruction at al-Basra is unclear in the chronicles, although future archeological work at the site may yield some clues.

During this period of conflict, the city of al-Basra was described by two Arab geographers who travelled through the Maghreb in the tenth century. Ibn Hawkal, a merchant and probable Fatimid sympathizer, journeyed through the area between 936 and 951 (Miquel 1971). He described al-Basra in this way:

> Al-Basra, a city of average extent, is enclosed by a weak wall. There is no water inside the city but it must be drawn at springs

> which flow in gardens situated outside the walls on the east side. The city produces a large quantity of cotton [flax?] which is exported to the province of Ifrikiya and elsewhere. The other products of its territory consist of wheat, barley and other cereals in large quantity. Abundance reigns at al-Basra; its bazaars are very fine, the air is pure, and the people are pious and studious. The inhabitants of both sexes are in general remarkable for their beauty and they have a fine and slender stature. They are distinguished by their modesty and their good manners [Ibn Hawkal 1842:192].

Ibn Hawkal provides an additional piece of information about al-Basra in a description of Tochmoch, a small city located along the Sagdad river (the present-day Loukkos River) about a mile from the ocean, perhaps near the contemporary city of Moulay bou-Selham.

> The Sagdad is composed of two branches, one of which rises in the mountains of al-Basra and crosses the land of the Denhadja. The other branch comes from the land of the Kutama. Both tributaries are very wide, and the inhabitants of al-Basra embark with merchandise on the branch that passes their city; they descend to the ocean in this way and then go whereever they wish. Tochmoch is about one day's journey away from al-Basra [Ibn Hawkal 1842:191].

The description of al-Basra on a tributary of the Sagdad is curious. The only river that flows past the site today is the Mda River, a small river that turns south toward Souk al-Arba of the Gharb rather than north toward the Loukkos River. The stream may have changed its course during the past millennium or the geographer, Ibn Hawkal, may have misconstrued the name of the city or the location of al-Basra.

The second geographer, al-Mukaddasi, traveled through the Maghreb at a slightly later time, sometime in the last half of the tenth century (Kramers 1936). Because of the brevity of his account of the Maghreb, it is conceivable that he did not observe firsthand many of the sights that he described. His reference to al-Basra is very short, but it provides an important piece of information: "al-Basra was a large city, flourishing and important, but now it is in ruins" (al-Mukaddasi 1950:273). This remark indicates that al-Basra was destroyed or badly damaged at the end of the tenth century.

## Zenata Confederacy and Almoravid Imperialism

After the Fatimids transferred their power base to Egypt, the Maghreb once again broke up into many small polities. Ifrikiya came under the authority of the Zirids, who acted as Fatimid vassals for a short time. Central Algeria was controlled by the Hammadids, who ruled from their new cities of Qal'a and Bougie. Northern Morocco came under the hegemony of the Spanish Umayyads and their allies, the Zenata Berbers. After the Umayyad Caliphate disintegrated in 1031, the Zenatas governed northern Morocco on their own accord as a loose tribal confederacy.

With the Almoravid invasion in the late eleventh century, the western Maghreb and Spain were again united under a single political authority. Under the charismatic leadership of Yusuf ibn Tashufin, the Almoravids, Berbers from the Sahara Desert, established the first Berber empire in the western Mediterranean. Centered in the new capital city of Marrakesh in southern Morocco, the empire controlled most of the trade routes between West Africa and Spain. Another invasion was underway in the eastern Maghreb at about the same time. The Hililian Arabs from the deserts of Egypt moved into the area at the instigation of the Fatimids, who wanted to revenge their former vassals, the Zirids. The Hililian Arabs gradually destroyed most of the cities in the region and, for the first time since the Arab invasions in the seventh century, began to assimilate the Berbers into Arab culture.

Just before the Almoravid invasion in the mid-eleventh century, the city of al-Basra was described by the Andalusian geographer, al-Bakri. Based entirely on secondary sources since he never traveled in North Africa himself, al-Bakri's accounts nevertheless are regarded as being remarkably precise (Lévi-Provencal 1960). His description of al-Basra suggests that the city had been rebuilt after it was dismantled in the tenth century:

> The city of al-Basra occupies a large site and surpasses all the neighboring localities in the extent of its pastures and the number of its flocks. There exists such an abundance of milk that the city has been called _al-Basra-t-ed-Dobban_ ("al-Basra of the Flies"). It is also called _al-Basra-t-el-Kittan_ ("al-Basra of the Flax") because when it first became populated, flax was used as money in all its commercial transactions. It is also called _al-Hamra_ ("the red") because the land on which it is built is a reddish color.
>
> Situated between two hills, this city is enclosed by a wall that is pierced by ten gates and constructed in stone and brick. There is a mosque with seven aisles and two baths. The principal cemetery is on a hill to the east of the city; the western cemetery is named _Macbera_ _Codaa_, ("the cemetery of the Arab tribe of Codaa").
>
> Since the water found in the city is brackish, the inhabitants get their drinking water at a well situated outside the principal gate named Bir ibn Delfa. The gardens outside of al-Basra contain numerous springs and wells that furnish sweet water.
>
> Adjoining al-Basra is a city of modern construction, which was founded at about the same time as Asila [al-Bakri 1965:216-217].

In another passage, al-Bakri notes that caravans from al-Basra, as well as from Fez and Sigilmasa, traded merchandise at a periodic market held at the fortress town of Hisn Day in southern Morocco (al-Bakri 1965: 294).

The possibility that al-Basra had been rebuilt is suggested by al-Bakri's remark that "adjoining al-Basra is a city of modern construction," built about the same time as Asila, which was founded by the Idrisids. As Chapter 4 shows, al-Bakri's comments are consistent with the archeological evidence from al-Basra. Structures from the final two occupation phases coexist side-by-side in some parts of the site.

Although al-Basra's existence in the middle of the eleventh century is substantiated by al-Bakri, its fate after this time is unclear. In the twelfth century, the geographer al-Idrisi claimed that the city was not as illustrious as it previously had been, implying--perhaps wrongly--that the city still existed in his time (al-Idrisi 1969:202). In the sixteenth century, Leo Africanus noted that only the walls of the formerly prosperous town remained (Leo Africanus 1956:259). In the nineteenth century, the French diplomat Charles Tissot wrote that he saw only rubble and debris scattered across the ground at the site of al-Basra (Tissot 1877:296).

As later chapters in this study show, the archeological evidence from al-Basra, in particular the pottery, indicates that the city was abandoned before or shortly after the Almoravid invasions in the late eleventh century. Examples of stamped and _sgraffito_ pottery associated with the Almoravid period at other sites in the western Maghreb and Spain have not been found in excavations conducted at al-Basra thus far.

## Economic and Sociocultural Aspects of al-Basra

As the chronicles indicate, al-Basra served as a small agricultural, commercial, and administrative center for most of its history. The city's economy was based primarily on agricultural production, in particular flax, wheat, and cattle. The city was surrounded by fertile agricultural land and pastures, much as the area appears today. Its early agricultural character is underscored by the nicknames attached to it: "al-Basra of the Flax" and "al-Basra of the Flies" (al-Bakri 1965). Its economy also involved the manufacture of craft goods, but to a lesser degree. According to one interpretation of the chronicles, the city may have been involved in the production of linen cloth (_mala_) (Eustache 1955:232). While this interpretation is speculative at present, it is possible that future archeological work at the site may unearth some evidence of cloth weaving. The city also produced pottery, as archeological data indicate, but most of it may have been made for local consumption.

According to the chronicles, the merchants of al-Basra shipped agricultural products to destinations in North Africa and Spain. They transported their goods by river to ports along the Atlantic, from which they shipped the merchandise by sea. In addition, they sent products overland to markets as far away as the south of Morocco (al-Bakri 1965).

The extent of al-Basra's involvement in long-distance trade between the Sahara and the Mediterranean is unknown, but it was obviously on a far smaller scale than that of Fez. Fez served as a major hub in the long-distance network, shipping goods from Sigilmasa east through the Taza corridor to Tlemcen, Tahert, and eventually to Kayrawan in Ifrikiya (Figures 3.1, 3.2). At most, al-Basra would have served as a stopover for caravans moving north from Fez to ports along the Atlantic and the Straits of Gibraltar, such as Larache, Tangier, and Ceuta.

According to the historical record, al-Basra's population consisted predominantly of Arabs and a small number of Berbers. The Berbers may have had a small settlement there before the Idrisids arrived in northern Morocco. At the beginning of the ninth century, the Arabs migrated to northern Morocco from Spain and other parts of the Maghreb. A number of Arabs came from Ifrikiya in about 805. In 818 several thousand newly converted Spanish Muslims (muwalladun) settled in Morocco and other parts of the Mediterranean after an unsuccessful revolt against the Spanish Umayyad government in Cordova (Dozy 1972:254; Lévi-Provencal 1944:119-120). The muwalladun consisted for the most part of farmers and craftsmen who undoubtedly brought their traditional agricultural and craft techniques to the Maghreb (Hole 1958; Lévi-Provencal 1944:120). In 826, another wave of immigrants arrived in Morocco, possibly stimulating some of the major construction projects that the Idrisids undertook a few years later.

## History of al-Basra within a Regional Context

Although the history of al-Basra is fairly sketchy, it is likely that its development and decline were influenced to a certain extent by regional political, economic and military events. For example, the establishment of the Arab city was closely tied to the foundation of the Idrisid state in northern Morocco. The reconstruction of al-Basra may be linked to the consolidation of the northern province under the authority of the family of al-Qasim, as well as to the arrival of political refugees from Spain. Its partial destruction may have been associated with the end of the Idrisid dynasty in Morocco and the withdrawal of the Fatimids from the Maghreb. Table 3.1 illustrates the continuities and discontinuities that existed between regional political changes and particular historical developments at the city of al-Basra itself.

In the first half century after its foundation (ca. 800-860), the city of al-Basra served as a small agricultural center in the centralized state that Idris II established at Fez but which, at his death, became territorially and politically divided. Al-Basra was assigned to the northern province, which had no permament ruler for several decades. Throughout this period the city continued to grow as new immigrants came into Morocco from Spain and other parts of North Africa.

During the next 120 years or so (ca. 860-980), the city became an important administrative and commercial center in the northern province. Sometime after 860, the province was consolidated under the authority of the Idrisid prince, al-Qasim, and his son, Ibrahim, who resided at al-Basra. The city's massive walls may have been built at this time. With a few exceptions, the family of al-Qasim governed the northern province from their various cities--al-Basra, Asila, and Hajar en-Nasr--for the duration of this period, at first independently, and after 917, as vassals of the Fatimids and the Spanish Umayyads. In 958 al-Basra became a provincial capital for the Fatimids in northern Morocco, and 20 years later it was partly destroyed.

TABLE 3.1

CHRONOLOGY OF MAJOR POLITICAL EVENTS IN MAGHREB
AND HISTORICAL DEVELOPMENT OF AL-BASRA

| Dates | Major Political Events in Maghreb | Historical Development of al-Basra |
|---|---|---|
| 800 | Formation of Idrisid State | Settlement of city by Idrisids |
| 850 | | |
| | | Idrisid provincial capital |
| 900 | | Construction of walls (?) |
| | | Rebuilding of city |
| 950 | Fatimid Empire | Destruction of city |
| | | Reconstruction of city |
| 1000 | Umayyad-Zenata Confederacy | |
| 1050 | | |
| 1100 | Almoravid Empire | Abandonment of city |
| 1150 | | |

For the next century or so, the reconstructed city served as a commercial-agricultural center under the decentralized Umayyad-Zenata confederacy in northern Morocco. Al-Basra seems to have disappeared completely sometime in the late eleventh or twelfth century.

# CHAPTER 4

# ARCHEOLOGICAL EXCAVATIONS AT AL-BASRA

The previous chapter examined the historical evidence for the foundation, florescence, and decline of the city of al-Basra. As it showed, the documentary record provides a sketchy and, in some cases, conflicting account of the city's past. This chapter and the next describe the archeological material uncovered at the thousand-year-old site during excavations in 1980 and 1981. The examination of the stratigraphy and the architectural and artifactual remains helps to amplify the documentary evidence and, in some cases, to resolve some of the problems raised by the historical record.

At the same time, the descriptive material presented in the next two chapters forms the empirical basis for the analytical and interpretive work that follows. Because of its importance to this body of research, the empirical data are discussed in some detail. This chapter provides an account of the sampling and excavation strategies, a description of the stratigraphy of individual soundings, a summary of major construction/occupation phases at al-Basra, and a brief discussion of the spatial layout of the urban center.

## Early Archeological Interest in al-Basra

Travelers have long expressed a curiosity about the ruins of al-Basra. In the late nineteenth century, M. Charles Tissot, Plenipotentiary Minister of France to Morocco, journeyed through the northern part of the country in order to locate and identify the classical sites in the region (Tissot 1877). According to Tissot, the ruins of al-Basra appeared to correspond with the location of the Roman town of Tremulae, which was supposed to have been situated about 18 km from Oppidum Novum, present-day Ksar el-Kebir (Figure 2.1). Tissot pointed to al-Basra's strategic position--dominating as it did the Mda River valley toward the west, a valley emptying into the Loukkos River basin toward the northeast, the road to Ouezzane toward the east, and the road from Ksar el-Kebir to Fez and Meknes toward the south--as an indication that the Islamic town must have been preceded by a Roman city. During Tissot's visit, however, only the remnants of the Islamic city wall remained. There was no evidence of a Roman town on the site, nor have recent excavations revealed any Roman material.

Archeological interest in the site emerged in the 1950s when M. A. Luquet, then Moroccan Inspector of Antiquities, conducted a survey for ancient sites along the Atlantic coast. His excavations at the coastal site of Moulay bou-Selham prompted speculation that the town had served as an entrepôt for the merchants of the city of al-Basra located 40 km to the east (Eustache 1955). Luquet's excavations at Moulay bou-Selham, however, showed, that the site dated to the twelfth-century Almohad period, not to the ninth and tenth centuries when al-Basra was a thriving commercial center (Eustache 1955). Luquet's findings were supported by Charles L. Redman's exploratory excavations at Moulay bou-

Selham in 1981, which indicated that the coastal site was probably not occupied prior to the twelfth century (Redman 1983, in press).

During a visit to al-Basra in 1955, Luquet drew a map of the site. Using this map, Eustache later published a description of the Islamic site (Eustache 1970-71). Because it describes the character of the site much as it appears today, Eustache's account is summarized below:

> The ruins of the small medieval city of al-Basra still exist today. After following the road that connects Souk al-Arba of the Gharb with Ksar el-Kebir for 9 km and taking the right hand turn to Ouezzane, the remains of the city are visible 13 km further down the road on the left side. The road crosses the low part of the ancient site, near the southern wall, of which only the foundations remain. The site is situated on the slope of a ridge that dominates the vast terraces of the Mda River valley.
>
> The city wall, of which only the northwest section is preserved, encircles the site for about 2.5 km. Its southwest corner is located several meters south of the road; the wall runs directly north from there, although only a line of asphodeles and agaves marks its foundations. After crossing a stream and entering a cactus thicket, the wall passes near an ancient cistern--still containing traces of its arches. Beyond the cistern, the first vestiges of the thick stone curtain which is supported by semicircular towers appear. The foundations of the wall continue for another 50 m, ending with traces of a possible gate. From this point, the wall remains intact for about 220 m, turning in an easterly direction two times and finally running along the ridge before ending abruptly. Eight semicircular towers support the walls that remain. Toward the north, a rectangular platform extends under the ridge, from which there is a beautiful view of the entire valley. Further east, inside the line of the wall, the remains of some structures are concealed in a cactus thicket. Throughout the site, the ground is littered with potsherds [Eustache 1970-71:118-119].

Eustache's description ends at this point, but the city wall continues its circuit around the site. From the north ridge, it drops down the hill to the road and then ascends the south hill, turning west along the ridge. Today, the contemporary village of Jouana-Basra is located on part of the north slope and villagers' gardens, orchards, and threshing fields extend across the valley floor.

Despite hopes that archeological work would begin at al-Basra in the 1950s in order to shed light on the city's ancient past, excavations did not start there for another quarter of a century.

## 1980-1981 Archeological Project at al-Basra

Archeological excavations began at al-Basra in 1980 and 1981 as part of the wider American-Moroccan Archeological Mission under the direction of Charles L. Redman (Redman 1983, in press; Redman et al. 1980). The long-term project, which had focused on the excavation of the Islamic-Portuguese port of Ksar es-Seghir on the Straits of Gibraltar since the mid-1970s, was aimed at testing a number of theories concerning the growth of cities and states in the Maghreb (Redman 1985). In order to accumulate comparative information, Redman conducted short-term excavations at several other important Islamic urban sites in northern Morocco. These were Nakur and Badis along the Mediterranean coast and al-Basra, Moulay bou-Selham, and Jebila along the Atlantic littoral.

Of these sites, al-Basra constitutes one of the earliest cities built by the Arabs in Morocco. In addition, it remains largely accessible to archeological work. Unlike many other early Arab cities in the Maghreb, such as Fez, Asila, and Tlemcen, the ancient site of al-Basra has not been built over by a modern city. The site of Nakur, another important early Islamic city that was investigated by Redman, also remained relatively untouched until the 1980s when it was covered by dam construction on the Nakur River.

## Site Sampling and Excavation Strategies

A multistage, stratified random sampling procedure was used to test the 30-ha area inside the city walls in order to assess the extent and depth of cultural remains below the surface and to collect a representative sample of material for analysis (Figure 4.1). During the 1980 field season, the entire site was sampled. It was divided into 10 zones based on modern land use and topographic features--such as cactus fields, village compound, and valley floor--and at least one sounding was placed at random in each zone. During the following field season, two zones (F and G) that had yielded the most promising material were investigated more thoroughly with additional soundings.

In addition to the stratified, random sampling strategy, the locations of several soundings were determined by the results of a magnetometer survey undertaken in the F zone where a pottery kiln had been found during the first season. Soundings were placed in areas that showed both high and low anomalies in the magnetometer results, partly to test the reliability of the method. Those soundings placed in areas with high, rectangular-shaped anomalies yielded a second kiln (F3) and the remains of a burned mud-brick structure (F5). Those placed in areas with low anomalies produced stone walls without any burned mud-brick (F2, F6) or no architecture at all (F4). These promising results suggested that the method might be useful in locating subsurface fired clay structures at the site in the future.

Finally, a couple of soundings were placed over architectural features that were visible on the surface of the site, including a plaster floor (G3) and a lime kiln (I2).

The excavation procedures were relatively straightforward. Soundings were generally 2-by-4 or 4-by-4 m in size, depending on their location. They were excavated horizontally by 20 cm. arbitrary levels, except when floors or other stratigraphic features occurred. They were separated vertically into loci, which were defined by stone walls or other architectural structures. The actual digging was conducted by locally hired workers who used picks and shovels for most of the work.

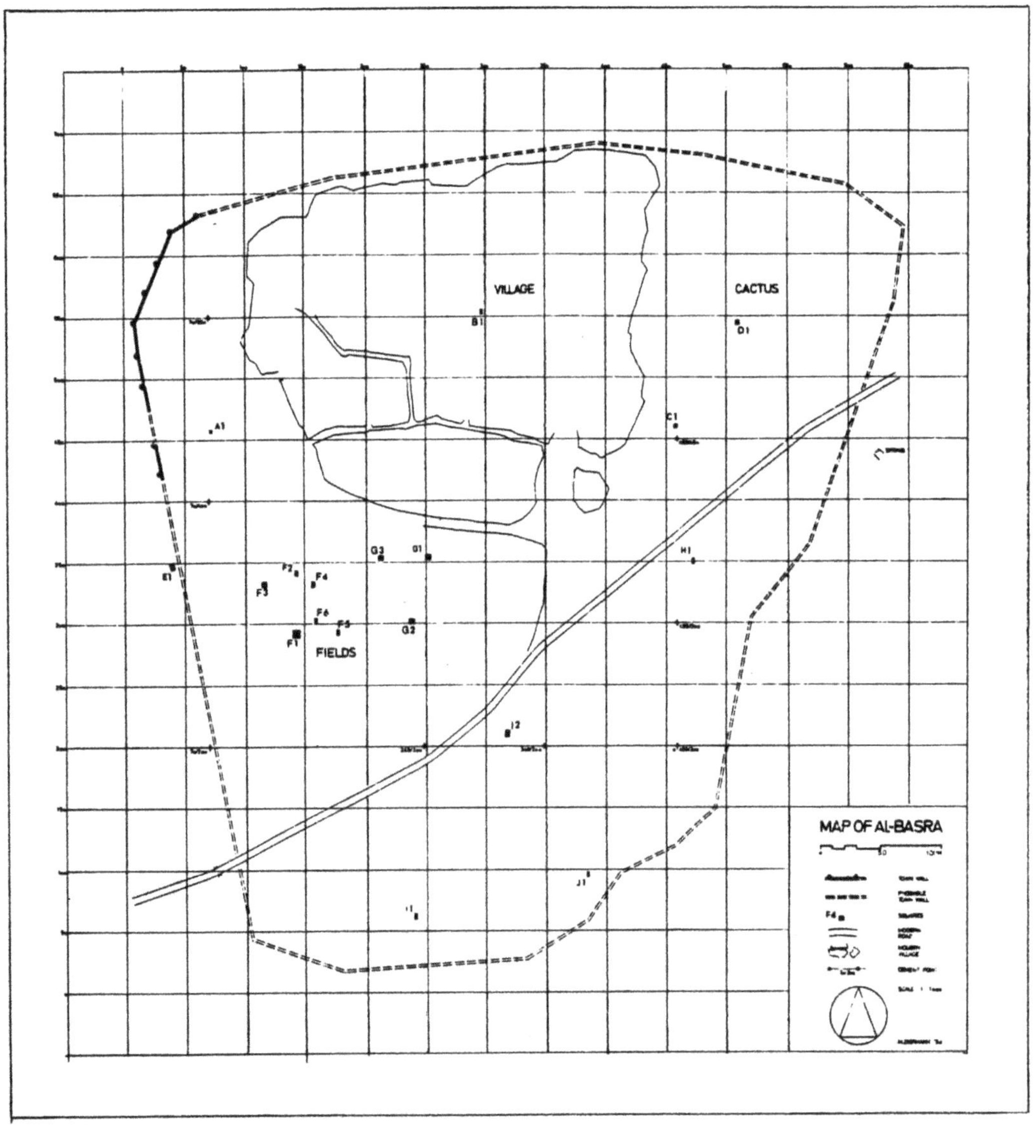


Figure 4.1. Soundings excavated at the site of al-Basra in 1980 and 1981.

They removed cultural material from the soil by hand. A few provenience units were screened to determine the amount of cultural material that was lost; this amounted to an estimated 10% by weight for pottery.

## Stratigraphic Descriptions

The stratigraphy of all 18 soundings excavated at al-Basra are described below. The descriptions are based on field notes, field drawings, and various published materials (Redman 1983, in press; Redman et al. 1980). The stratigraphy of some of the more important soundings are illustrated in accompanying sections and plans (Figures 4.2-4.17).

### Residential Area (G Soundings)

Three soundings that were located near the center of the site in the lowest part of the valley revealed what appears to be the residential area of the city. Two of the units yielded a series of superimposed stone walls, which extended 3 m below the surface. These units provide a fairly comprehensive picture of the successive building and/or occupation phases at the site.

#### G1 Sounding

G1 (Figure 4.2), a 4-by-4-m excavation unit located about 20 m south of the village compound, yielded a long stratigraphic sequence featuring at least three building levels.

The uppermost 50 cm consisted of dark brown topsoil that accumulated as a result of wind and water erosion of the hills around the site; this deposit contained relatively little cultural material. The remnants of two roughly built stone walls, one running north-south and the other east-west (Wall 2), were found toward the bottom of the deposit. The next 50 to 60 cm consisted of burned red clay that contained a large quantity of broken roof tiles and relatively little pottery or bone. The red clay deposit, which was also evident in many other soundings at the site, consisted of disintegrated mud-brick walls and collapsed tiled roofs from structures apparently destroyed by fire.

Two limestone-and-mortar walls, one oriented east-west (Wall 3) and the other north-south, appeared below the red clay. The extant walls measured about a meter high and half a meter wide; a 20 cm wide channel ran down the middle of the north-south wall and across the width of the east-west wall at the point where they joined together.

In the northeast quadrant of the sounding were 80 cm of greenish-brown fill containing a moderate amount of pottery and bone. South of Wall 3 was a limestone deposit, possibly constituting the remains of a plaster floor. Below the limestone were 70 cm of fill, similar to that present in the northeast section, extending down to nearly the base of the walls. A drain-like structure was found a meter south of Wall 3.

The 20-30 cm directly below the walls contained a large quantity of stone rubble, pottery and bone. This deposit may have consisted of construction fill because of its high density of cultural material.

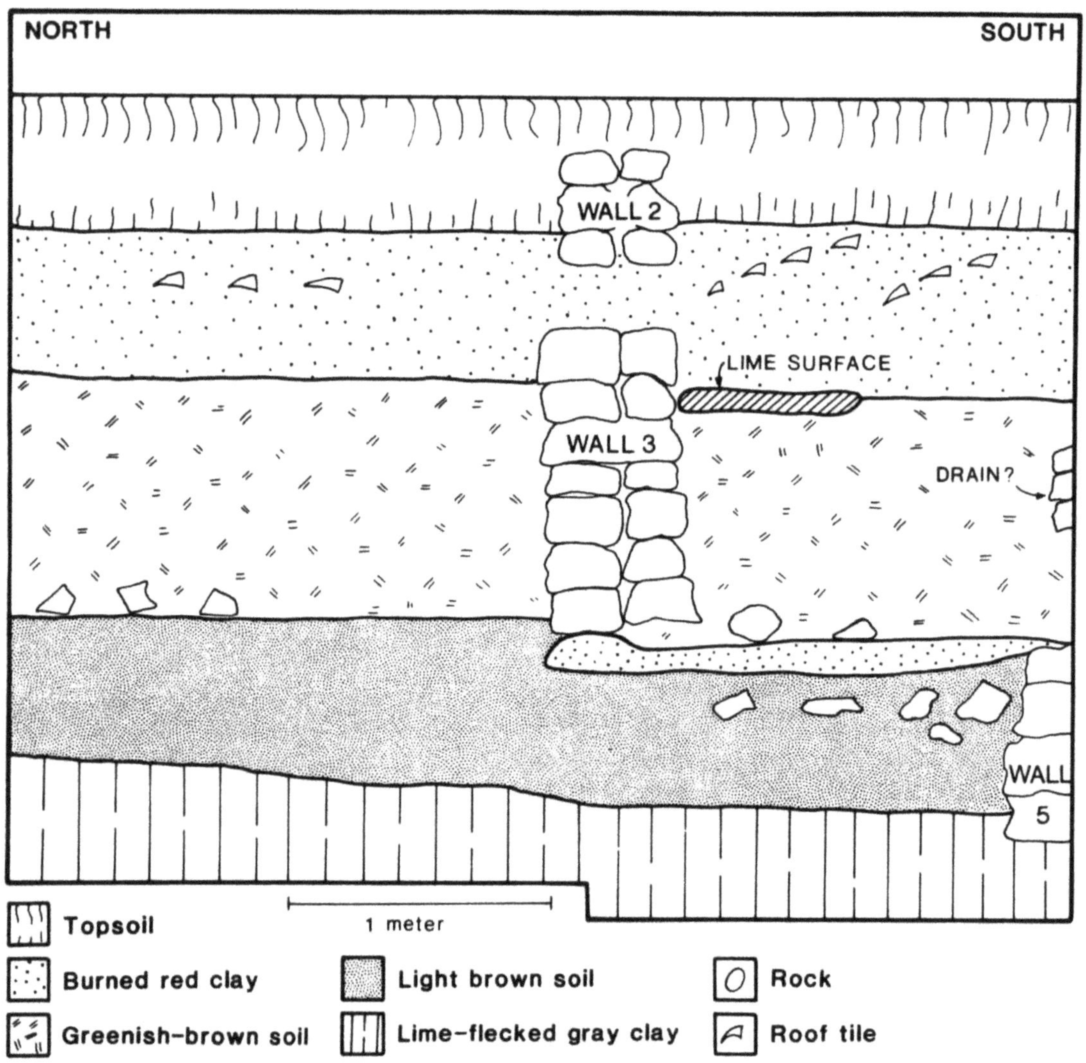


Figure 4.2. East section of G1 sounding.

Another stone wall (Wall 5) running in an east-west direction appeared below the fill. The wall was covered with 20-30 cm of burned red clay representing some sort of burning episode; below the red clay were 30 cm of fill containing a high density of bone and pottery. The lower wall stood in a lime-flecked gray clay that was culturally sterile.

G2 Sounding

G2 (Figure 4.3), a 4-by-4-m excavation unit located about 50 m south of G1, also revealed a deep stratigraphic sequence featuring at least three construction phases.

The uppermost 40-50 cm consisted of dark topsoil with very little cultural material. The next 40-90 cm consisted of burned red clay that contained large tile fragments, presumably from a collapsed roof. The red clay stratum rested on top of two roughly built stone walls (Walls 1 and 2), which ran parallel in an east-west direction about 60 cm apart. The walls were built of large, unshaped rocks that were mortared with mud, bone, and sherds, and rested on a cobble foundation. Large pottery fragments, probably the remains of vessels broken _in situ_ when the roof fell, were found in the narrow passageway between the walls. South of the walls below the red clay layer, a basin-like structure built of large flat rocks rested on a pebbled floor (Figure 4.4). Several broken storage jars were found _in situ_ on the floor near the basin, indicating that the area--possibly an outdoor courtyard--was abandoned quickly, perhaps just before the buildings collapsed. Underneath the pebbled floor were 20 cm of red clay and below that 30 cm of greenish-brown fill containing numerous pebbles.

In the southwest quadrant of the sounding, the red clay rested on top of two well-built limestone and mortar walls (Walls 3 and 4), which were oriented in a north-south and east-west direction and joined together at an acute angle. The walls stood about 1 m high and 52 cm wide and consisted of seven courses of stone. The lower two courses jutted outward from the interior face of the wall, possibly to support the weight of a 10 cm thick plaster floor that extended southward from the walls. The floor rested directly below the red clay deposit inside the walls. The floor also extended outside the walls where it appeared to have been blackened by fire. The limestone and mortar walls were supported by a foundation trench that was partly filled with an ashy residue.

Below the limestone walls were 20-30 cm of fill that contained a large quantity of stone rubble, bone and pottery, similar to that present below the limestone walls in the G1 sounding.

The next 20-30 cm consisted of burned red clay with large tile fragments scattered throughout it. The clay deposit rested on top of another wall (Wall 5), which ran in an east-west direction. A flagstone pavement extended southward from the base of this wall. A greenish-brown fill associated with the wall contained relatively little cultural material. The wall stood on a gray sandy soil below which was a culturally sterile gray clay.

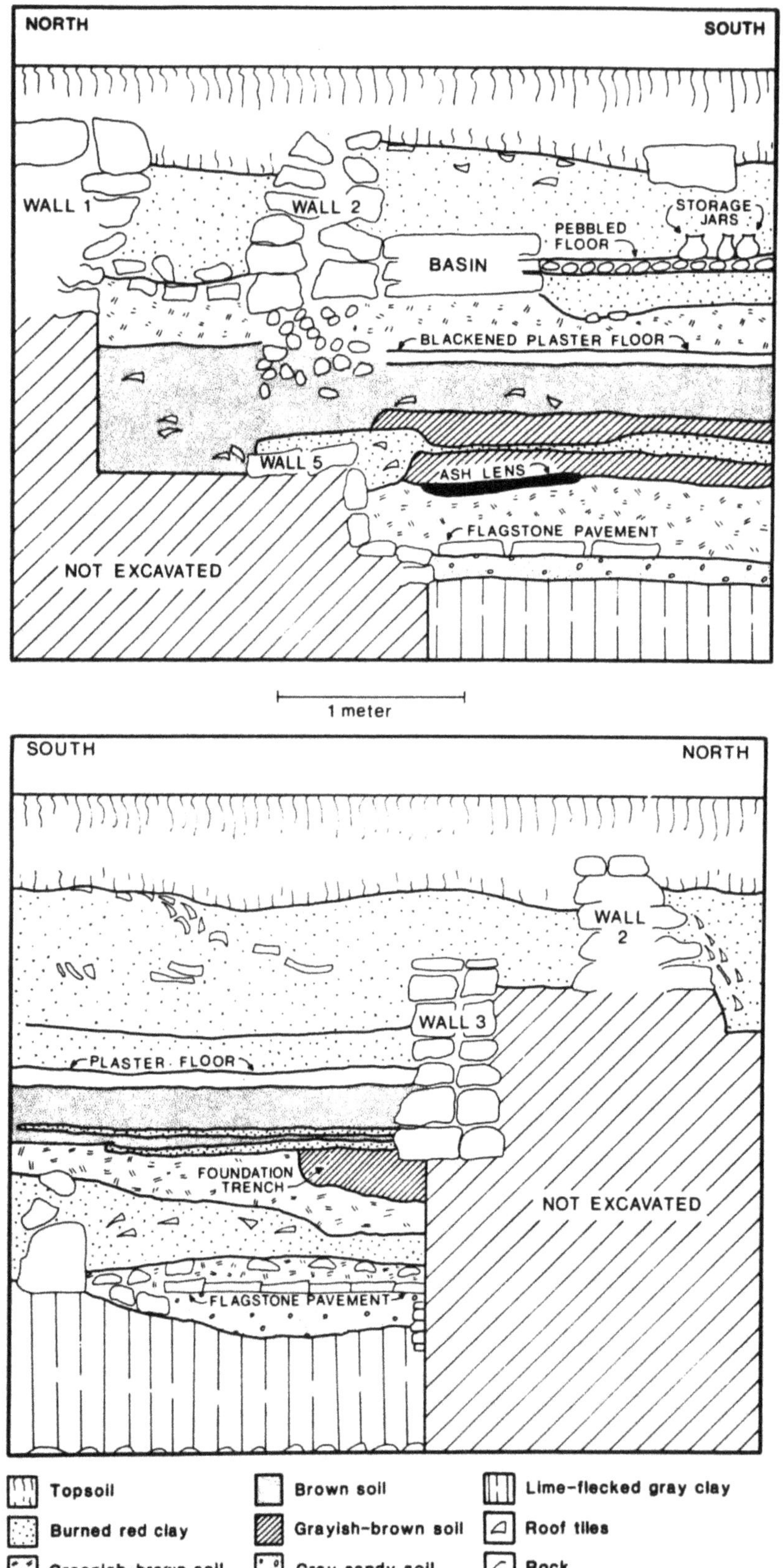


Figure 4.3. East (top) and west (bottom) sections of G2 sounding.

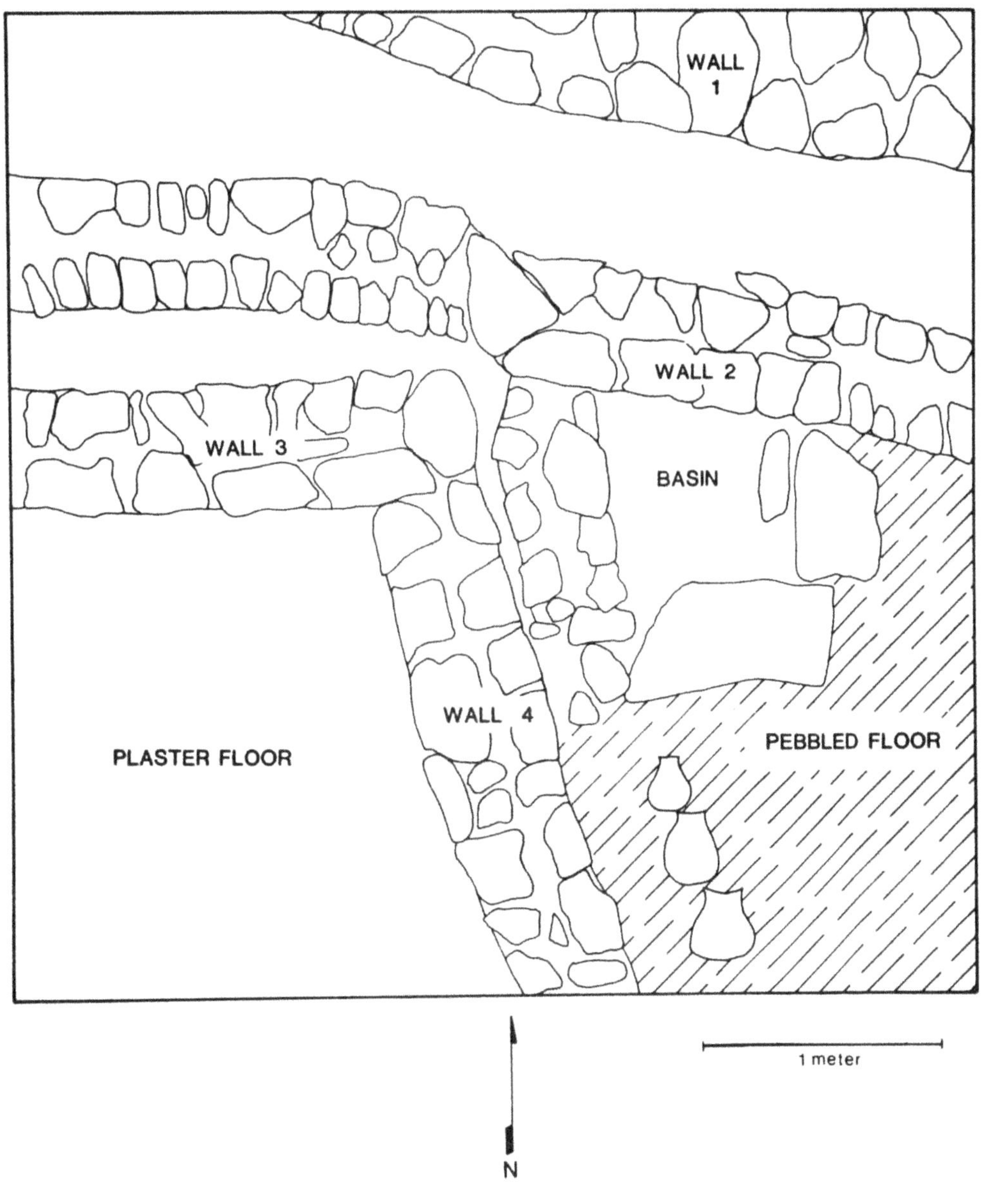


Figure 4.4. Plan of stone basin and pebbled floor, with associated storage jars and an earlier structure with a plaster floor, in G2 sounding.

G3 Sounding

G3 (Figure 4.5), a 6-by-8-m sounding located about 40 m west of G1, was placed over a plaster floor that was partly exposed by a villager planting his garden. Only the top 50 cm or so of fill below the surface of the ground was excavated due to time constraints.

The uppermost 20-30 cm consisted of dark topsoil. The stubs of a roughly built stone wall (Wall 1) that ran in a north-south direction appeared toward the bottom of the deposit.

The next 20-30 cm consisted of burned red clay that contained abundant roof tile and plaster fragments. This material rested directly on top of a nearly intact plaster floor that apparently belonged to at least two rectangular-shaped rooms. The southernmost room, which measured about 2 by 4.5 m in size, was connected by a narrow corridor to a second room, which was 2 by 4 m in size. A partially excavated corridor extended northward from this room. The floors were made of a 10 cm thick mixture of plaster and stone and were partly decorated with red paint. They were curbed at the point where they joined the walls of the building. A silver coin was found on the floor of one room, suggesting that the dwelling may have been abandoned abruptly. Unfortunately, the coin was badly abraded and thus has not been dated.

It is noteworthy that the layout of the rooms--with their connecting corridors--is significantly different from the traditional arrangement of Islamic houses, in which rooms lead directly onto a central courtyard rather than into another room. Examples of houses built around a courtyard were found in fourteenth- and fifteenth-century Islamic levels at Ksar es-Seghir (Jeffers 1981). They also appear in contemporary villages in the Gharb and Rif areas of northern Morocco (Mikesell 1961).

<u>Industrial Area (F Soundings)</u>

Six soundings situated between the residential area and the western wall of the city produced evidence of industrial activities, including the remains of two updraft pottery kilns. The soundings generally contained less architecture and larger quantities of pottery and bone than the residential area, suggesting that the area may also have been used as a dump for city trash.

F1 Sounding

F1 (Figure 4.6), a 6-by-6-m sounding located about 80 m east of the city wall, produced the remains of an updraft ceramic kiln below a meter of relatively undifferentiated fill.

The uppermost 50 cm consisted of dark topsoil that contained a greater quantity of small, worn potsherds, bone, and pebbles toward the bottom of the deposit. The abraded condition of the pottery suggested that the material had been transported for some distance.

The next 20-30 cm of grayish brown soil contained moderate amounts of pottery and bone. Below this was 20 to 70 cm of light brown sandy

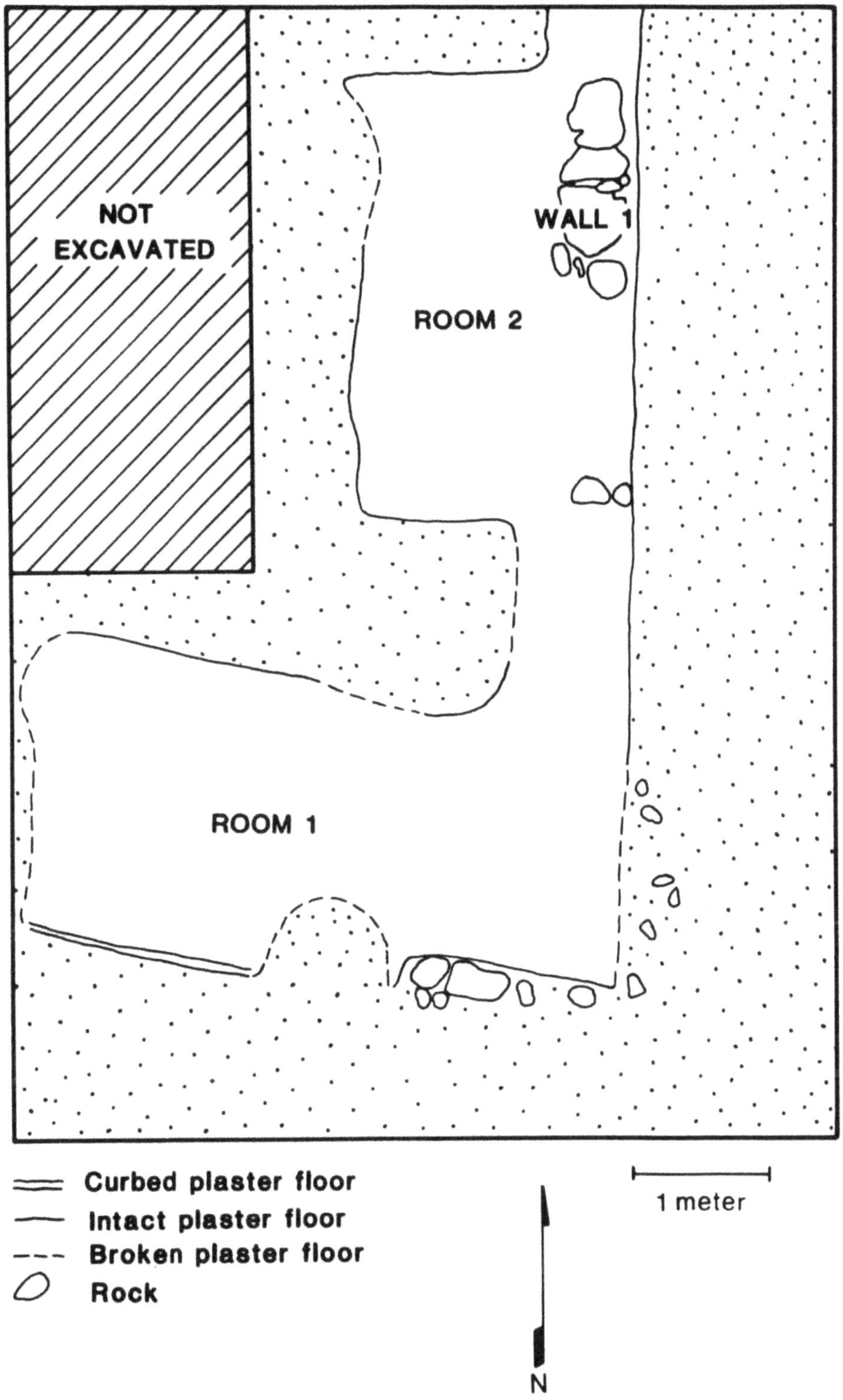


Figure 4.5. Plan of plaster floors in G3 sounding.

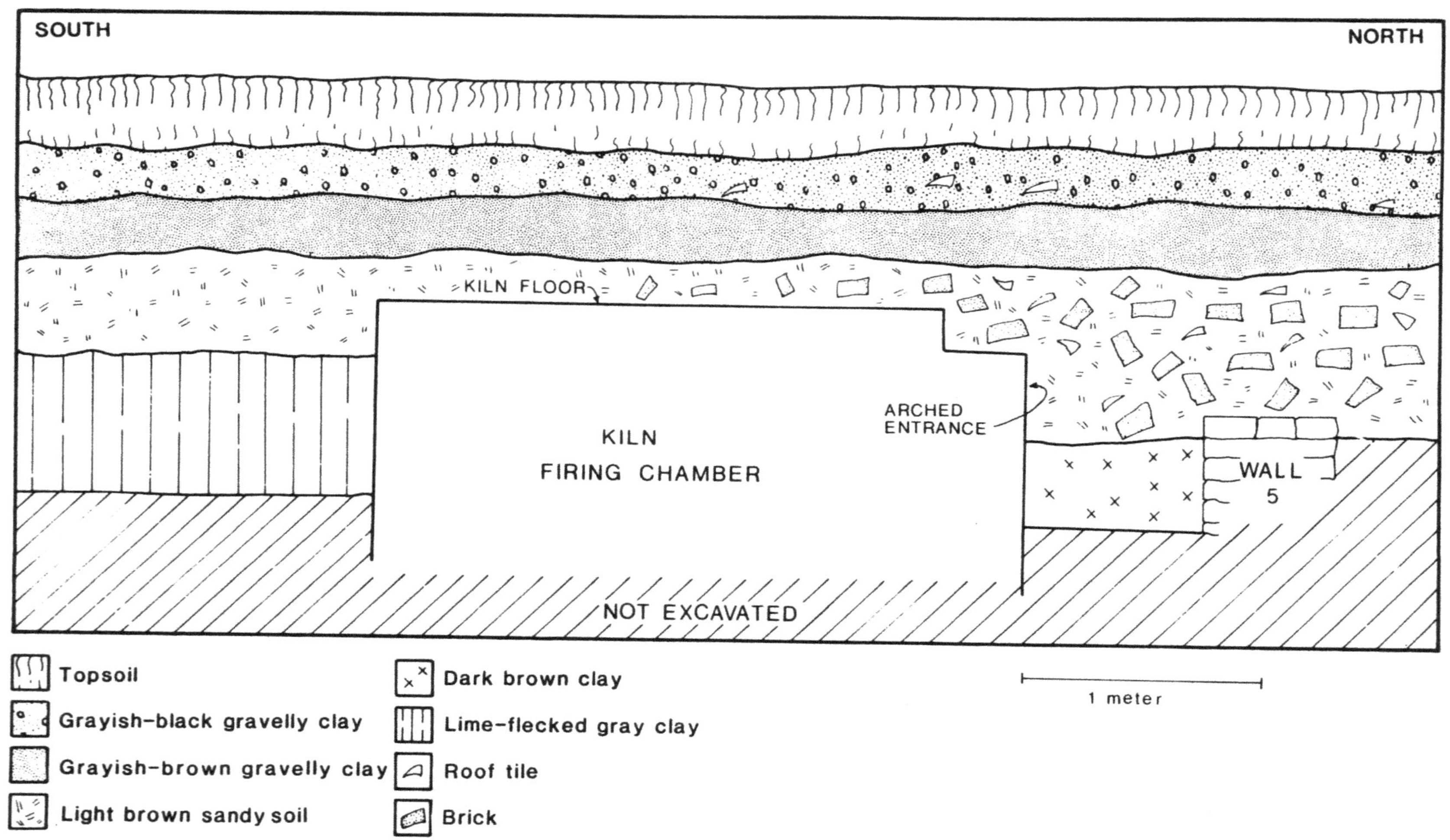


Figure 4.6. West section of F1 sounding.

fill that became increasingly clayey toward the bottom. The fill contained fragments of broken brick and tiles from the kiln collapse.

Below the brick rubble was the floor of the firing chamber, which measured 3.2 m long and 2.4 m wide and was made of fired brick lined with plaster. The floor was supported from below by a series of transverse arches made of brick. It was lined with several rows of flue holes that allowed heat to rise from the firebox up into the chamber where the pottery was stacked. The entrance to the firebox consisted of an arched doorway built of brick and mortared with tiles. The form of the upper chamber is not clear. It is possible that it was not a permanent structure because relatively little broken brick was found in the excavation unit. If it were not a permanent structure, the chamber may have been sealed with tiles or mud during firings, as it is done in many Islamic areas today (Rye 1981; Rhodes 1981). This is possible because of the unglazed nature of the al-Basra pottery, which would not have required the high firing temperatures that a permanently enclosed upper chamber could provide. In addition, much of the pottery is characterized by buff-colored surfaces around a gray core, indicating that it had been cooled abruptly in an oxidizing atmosphere at the end of firing, a process that would have probably cracked a permanently enclosed kiln (Rye 1981:110). North of the kiln was a brick wall that ran parallel to it. The space between the wall and kiln was filled with brick rubble, suggesting that the floor may originally have extended to the wall and later collapsed (Redman 1983). If this were the case, the original kiln would have measured about 4.75 by 4.0 m in size.

To the west and south of the kiln, the topsoil extended down to a depth of half a meter. Below this was half a meter of fill that contained a high density of bone and pottery. Some of this material may have consisted of discarded pottery from the operation of the kiln. Below the fill was a gray calcareous clay with little cultural material, indicating that the lower part of the kiln had been built into sterile ground.

F3 Sounding

F3 (Figure 4.7), a 6-by-5-m excavation unit located about 50 m northwest of F1, yielded the remains of a second updraft kiln. Unlike F1, however, the F3 kiln was overlain by at least two subsequent building phases.

The uppermost 30 cm consisted of dark brown topsoil with little cultural material. Below this were 30 cm of burned red clay that contained numerous roof tiles. Underneath the red clay were the remnants of a gravel surface and a stone wall (Wall 1) oriented in a north-south direction. The associated fill contained a large quantity of stone, pottery, and bone. A few meters east of the wall was an unusual feature consisting of a circle of stone rocks about 80 cm in diameter, whose function is unknown.

A well-built limestone-and-mortar wall (Wall 2), running in an east-west direction, was found south of Wall 1, but at a lower elevation. The wall, which measured about 70 cm high and half a meter wide,

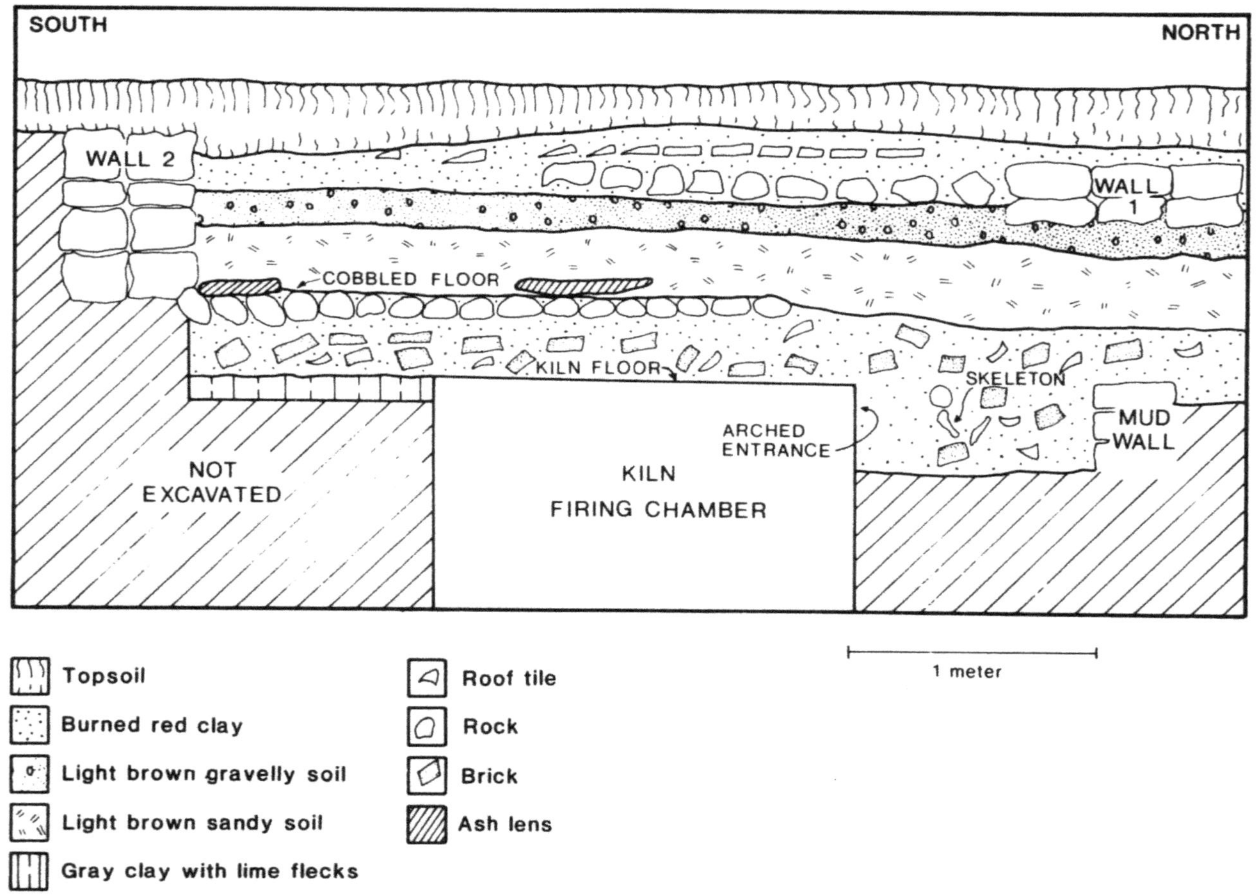


Figure 4.7. West section of F3 sounding.

was associated with a brown sandy fill that contained low quantities of bone and pottery, although the sherds were relatively large in size. A cobbled floor extended north from the foot of the wall, covering an area of about 2.5 by 1.5 m.

Below the cobbled floor were 20 cm of burned brick and clay containing numerous roof tile fragments. The deposit rested directly above a kiln floor, which measured about 3.25 by 1.8 m. Similar to that in the F1 kiln, the floor was lined with several rows of flue holes and was supported from below by a series of transverse brick arches. There was also a brick wall running parallel to the kiln about one meter north, which may have formed the original outside wall of the kiln. Parts of a human skeleton were found in the rubble between the kiln and brick wall. Redman has proposed the individual may have died in the kiln when it collapsed or the bones of a dead person may have been thrown into the kiln when it was no longer in operation (1983).

Deposits south of the kiln below the level of the kiln floor consisted of a culturally sterile gray clay, indicating that the kiln had been built directly into sterile ground.

Very little pottery was found inside either kiln. It is likely that broken or overfired vessels were discarded in nearby waster dumps rather than being allowed to accumulate on the floors or in the firing chambers where they would have interfered with the operation of the kilns.

F5 Sounding

F5 (Figure 4.8), a 2-by-4-m sounding located midway between the kiln and residential areas, produced a relatively deep stratigraphic sequence with some architectural features.

The uppermost 50 cm consisted of dark topsoil that contained very little bone or pottery. The next 30-40 cm consisted of burned red clay with numerous tile fragments. This material rested on top of a stone wall (Wall 1) that ran in an east-west direction. Constructed of large, unshaped stones mortared with tiles, sherds and bone, the wall stood about one meter high on a base of river cobbles.

Below the red clay north of the wall was a plaster surface that may represent the remains of a floor. The next 50 cm consisted of loose, brown fill that contained a large quantity of cultural material. Below the red clay south of the wall was a pebbled surface underlain by another 20 cm of red clay material. The next 20 cm consisted of a grayish-brown deposit with an ash lens at the bottom; this fill contained a relatively large quantity of cultural material, particularly burned animal bone.

The 60 cm deposit below the wall consisted of a sandy soil that became increasingly gravelly toward the bottom; below this were 20 cm of dark brown sandy fill. Both deposits contained moderate quantities of bone and pottery. The last meter of deposit consisted of gray clay flecked with limestone and contained very little cultural material.

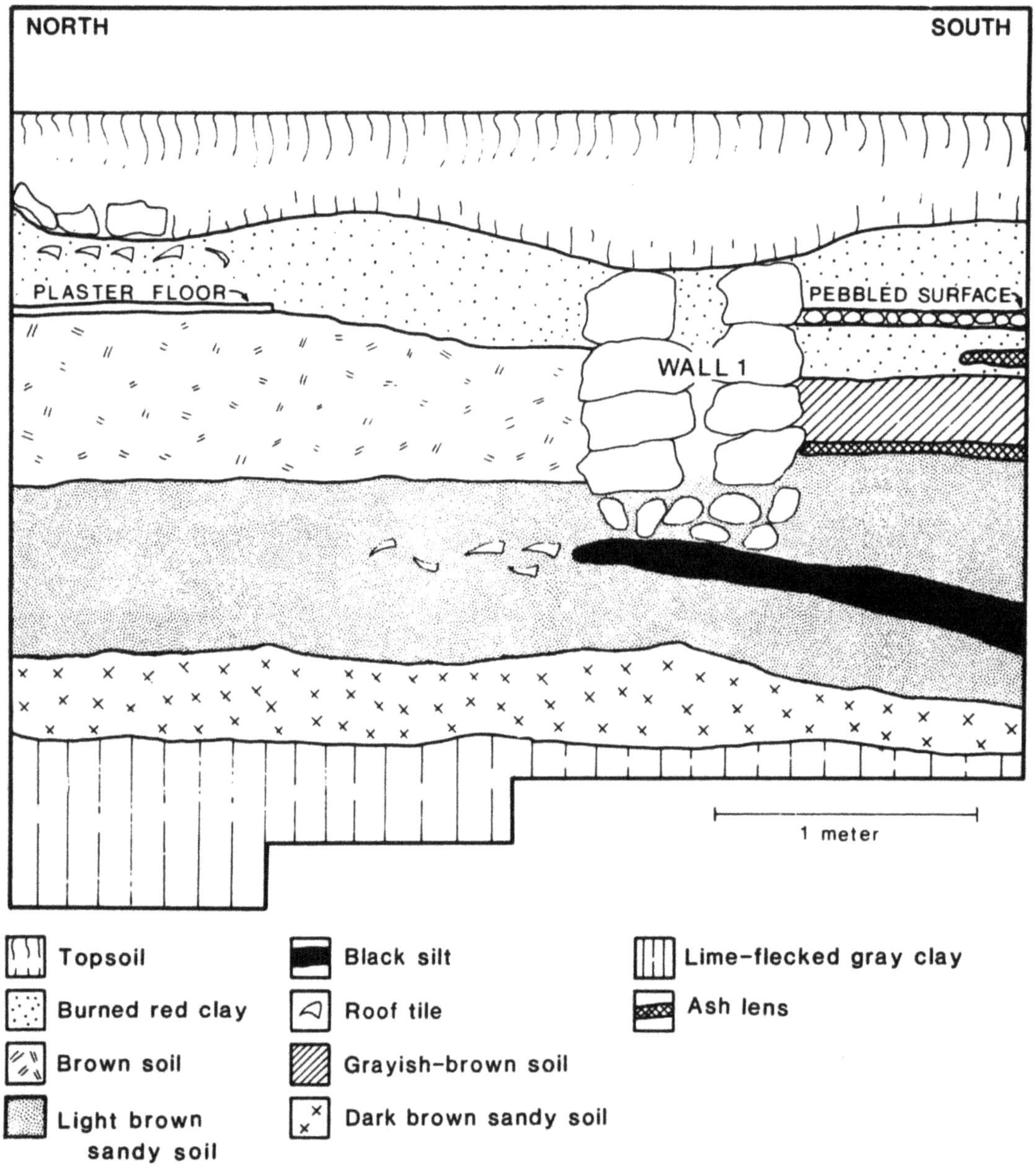


Figure 4.8. East section of F5 sounding.

F2, F4, and F6 Soundings

F2, F4, and F6, all 2-by-4-m soundings located near the pottery kilns, yielded relatively large quantities of cultural material and few architectural features. Excavation of all three soundings stopped before sterile ground was reached because the field season came to a close.

In F2 (Figure 4.9), the uppermost 40 cm consisted of black topsoil, followed by 20 cm of brown silt. The remnants of two walls (Walls 1 and 2), which ran parallel to each other in an east-west direction, were found toward the bottom of this deposit. The base of one wall ended at this level while the other extended down another meter or so. The next 50 cm consisted of a yellowish-brown sandy fill that graded into an increasingly gravelly deposit containing few artifacts. The number of sherds and bone increased in the sandy deposit below the gravel, suggesting that cultural deposits continued below this level.

In F4 (Figure 4.10), the uppermost 50 cm consisted of topsoil containing little cultural material. The next 50 cm consisted of a large quantity of gravel and a moderate amount of bone and sherds. Below the gravel was a light sandy fill that extended down another 80 cm. The fill contained a large quantity of cultural material, but no architectural features.

In F6 (Figure 4.11), the uppermost 60 cm consisted of topsoil. Toward the bottom of this deposit were two walls--the remains of a roughly built stone wall (Wall 2) and a nearly intact limestone and mortar wall (Wall 1)--which were oriented in an east-west direction. To the south of Wall 2 was a pebbled surface, possibly a floor, which was covered with wall rubble. Below the pebbled surface were 30 cm of brown fill. To the south of Wall 1 was a concentration of roof tiles. The wall, which was built of limestone and mortar, stood about a meter high and half a meter wide. The 60 cm deep deposit associated with the wall consisted of a loose, sandy fill that became increasingly gravelly toward the bottom; the deposit contained a very large quantity of pottery, along with bone and roof tiles. Below the base of the wall was another 30 cm of sandy fill that contained moderate amounts of cultural material and very little gravel.

City Water Cistern

E1 (Figures 4.12-4.13) was a 4-by-5-m sounding placed over the traces of a large cistern located directly outside the western wall of the city. The cistern was apparently the same structure that Luquet and Eustache described in the 1950s (Eustache 1970-71).

Inside the cistern the uppermost meter of fill consisted of wall and roof rubble, including tile fragments and large rocks. Below this were 60 cm of dark, sandy fill that contained fragments of unusually thick, heavy pottery vessels; similar sherds were also found in other soundings at the site, suggesting that they may not have been used specifically for cistern-related activities. The final 40 cm of excavated fill consisted of a yellowish sandy soil that contained large rocks but very little bone or pottery.

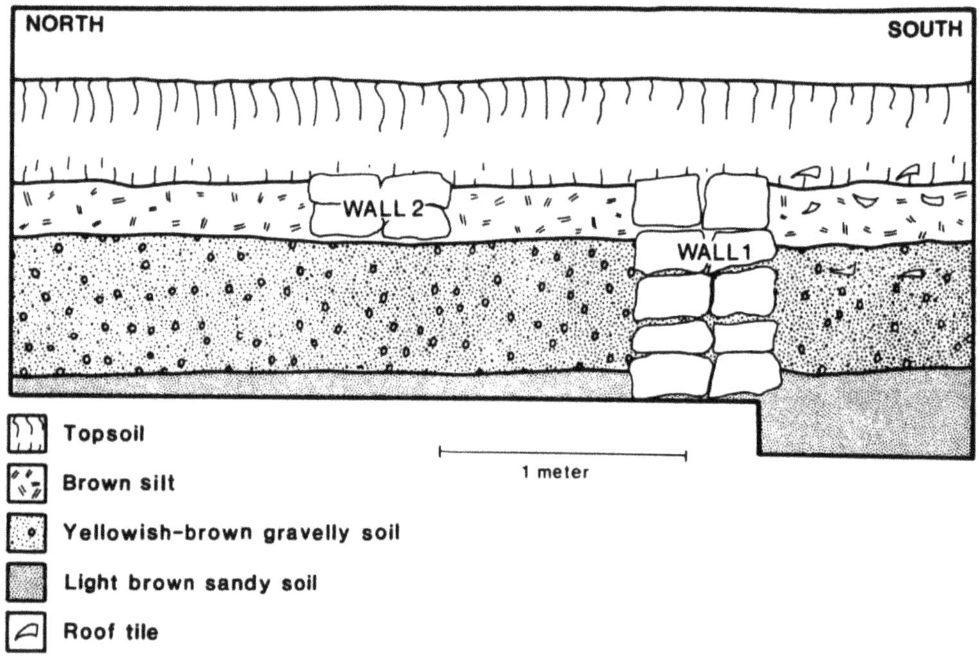


Figure 4.9. East section of F2 sounding.

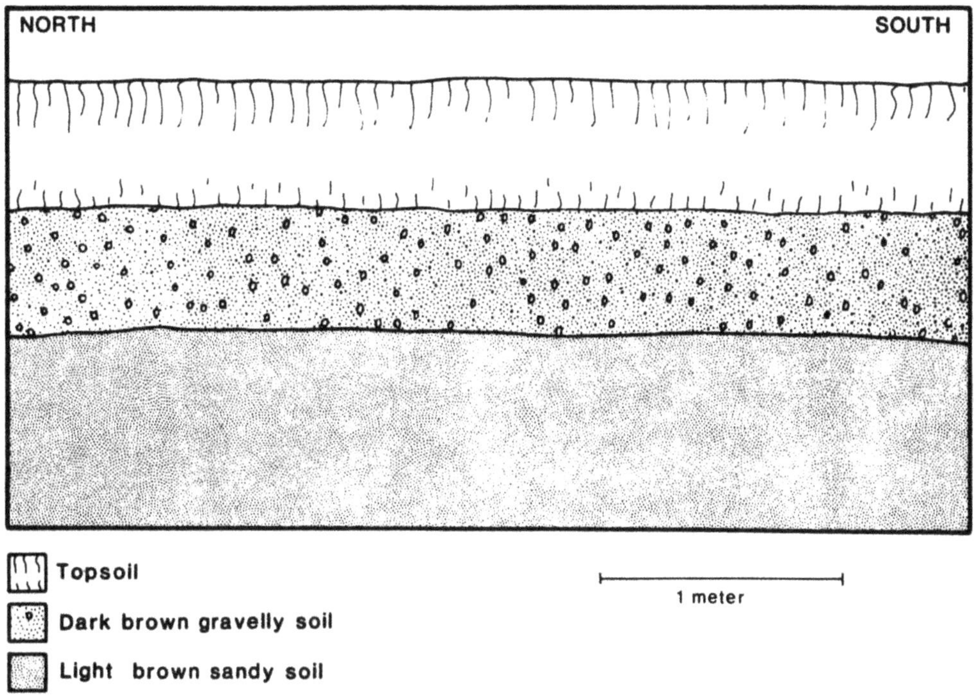


Figure 4.10. East section of F4 sounding.

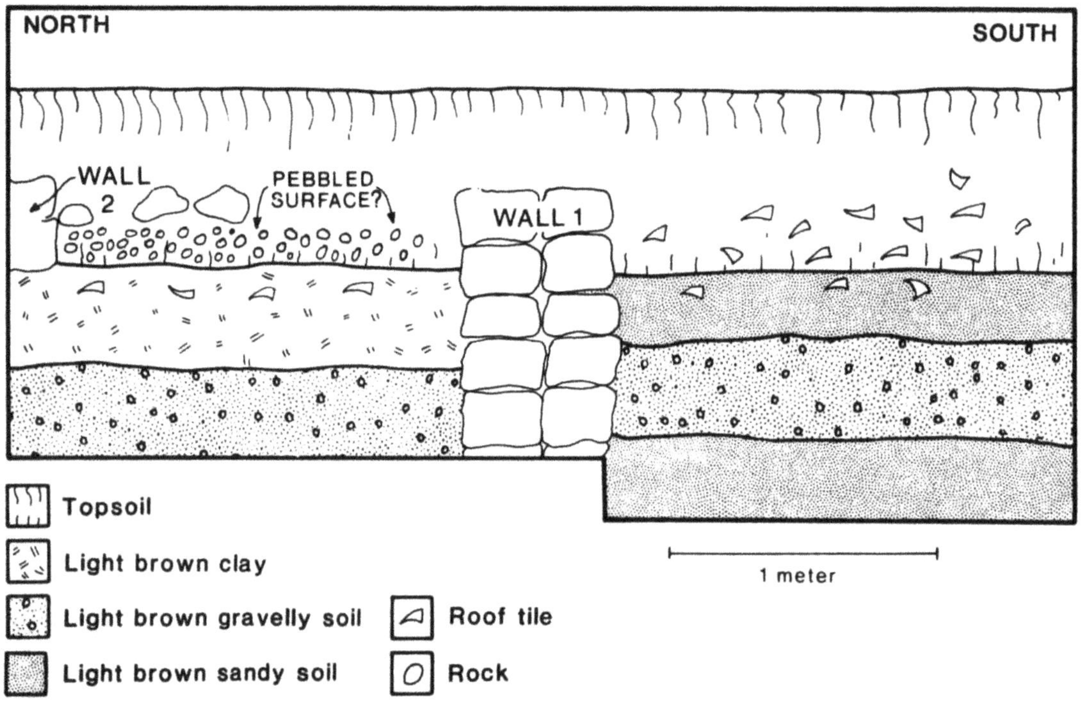


Figure 4.11. East section of F6 sounding.

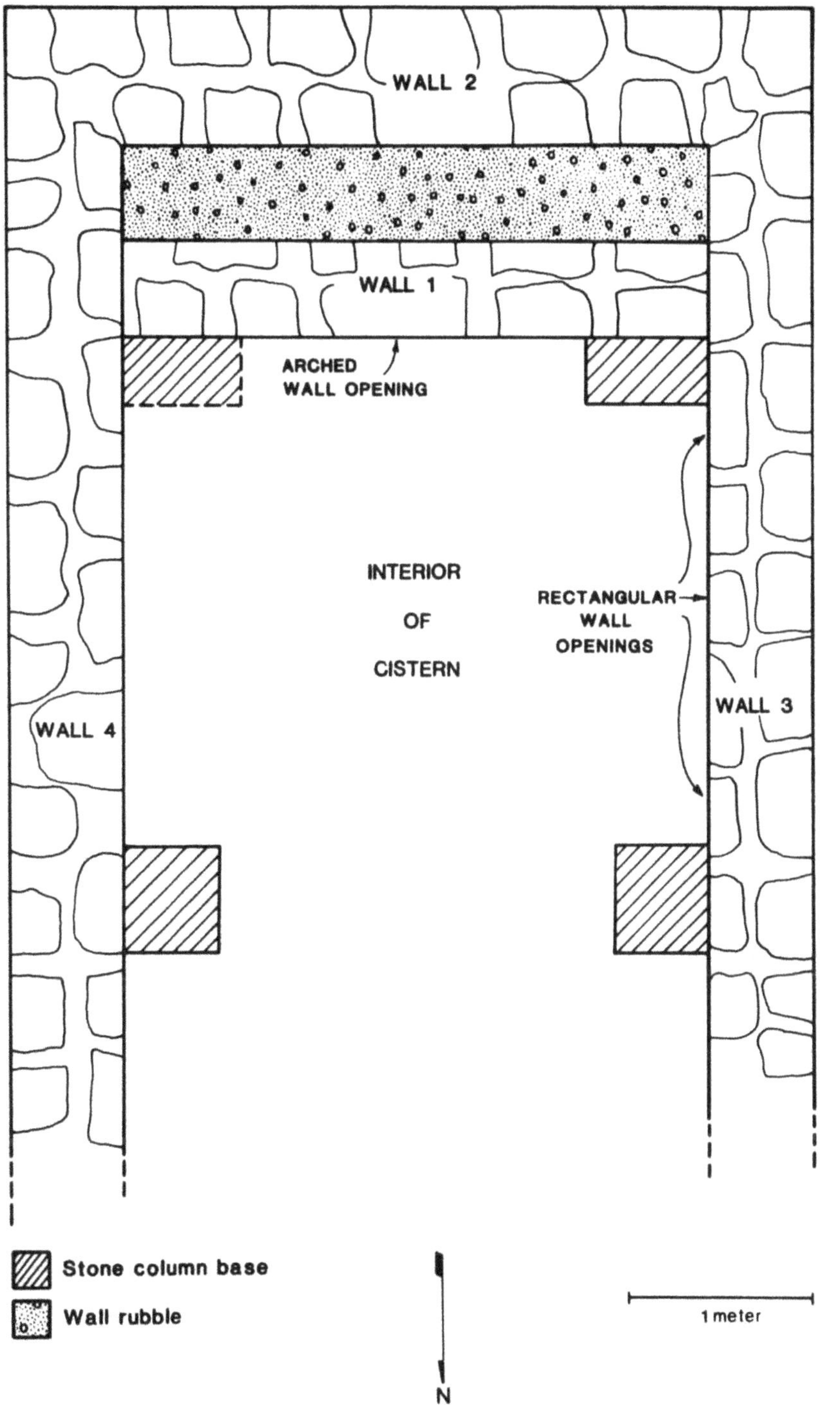


Figure 4.12. Plan of cistern in El sounding.

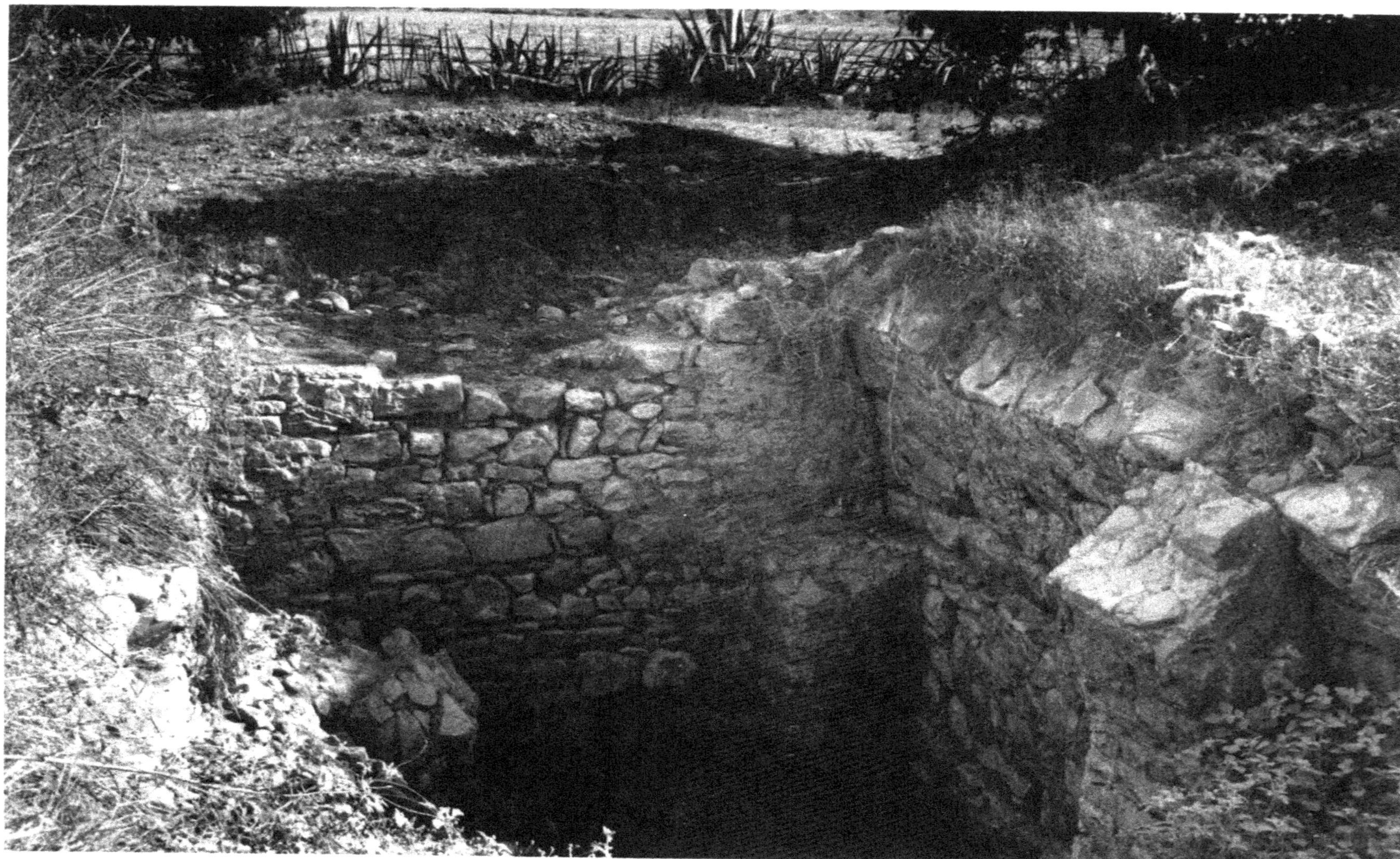

Figure 4.13. View of cistern in E1 sounding, looking south.

The excavated portion of the cistern measured 4.25 m wide and more than 6 m long on the exterior and at least 3 m deep. The walls were constructed of limestone rocks and mortar. In each corner of the cistern were stone bases to support transverse arches that had once held up a vaulted roof. Toward the bottom of the cistern on the west wall were small, rectangular openings that apparently allowed water to flow in and out of the system.

Lime Kiln

I2, a 4-by-6-m sounding located near the main road on the south slope, revealed the remains of an open-air lime kiln on top of several stone walls.

Below a thin band of topsoil, a thick layer of burned lime extended across an area of about 3 m in diameter and up to one meter deep. The deposit consisted of distinct layers of lime, indicating a succession of firings. The absence of any architecture associated directly with the lime deposit suggested that it may have operated as an open-air kiln. Below the lime deposits were the remains of several stone walls, generally running in a north-south direction. The walls were covered with a layer of burned red clay that contained broken roof tiles; the fill associated with the walls contained a small amount of pottery and virtually no bone.

Minor Soundings

Several soundings placed on the north and south slopes of the site produced very little cultural material or architecture. These shallow soundings reached sterile ground at less than one meter in depth.

A1, a 2-by-4-m excavation unit located on the north slope west of the village, revealed the remains of a stone wall oriented in a north-south direction. The wall stubs were found at the bottom of 50 cm of topsoil. Below this were 50 cm of fill that contained few artifacts and became almost entirely gravel at one meter in depth.

B1 (Figure 4.14), a 2-by-4-m sounding located in the village compound on the north slope, yielded no architectural features. Below 15-20 cm of topsoil was a layer of pebbles embedded in a red clay matrix that appeared to constitute bedrock on the hill slopes. This deposit contained very little pottery and bone and no roof tiles.

C1 (Figure 4.15), a 2-by-4-m unit located east of the village compound on the north slope, produced the remains of a brick structure. Directly below the surface were 50 cm of burned red clay that contained numerous brick, tile, and plaster fragments. The material rested on top of a layer of large rocks, possibly the remains of a wall. The next 40 cm consisted of a sandy soil that became increasingly gravelly toward the bottom and contained no cultural material.

D1 (Figure 4.16), a 3-by-3-m sounding situated in a cactus field east of the village, yielded the remains of a possible structure. Below 30 cm of topsoil was a deposit that contained plaster, tile, and pottery fragments, perhaps representing the collapse of a building.

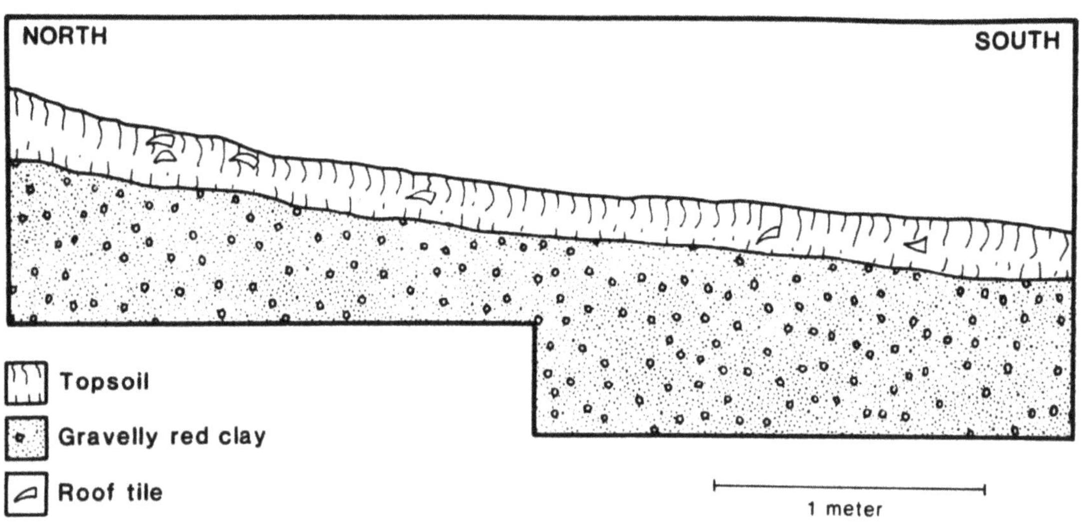


Figure 4.14. East section of B1 sounding.

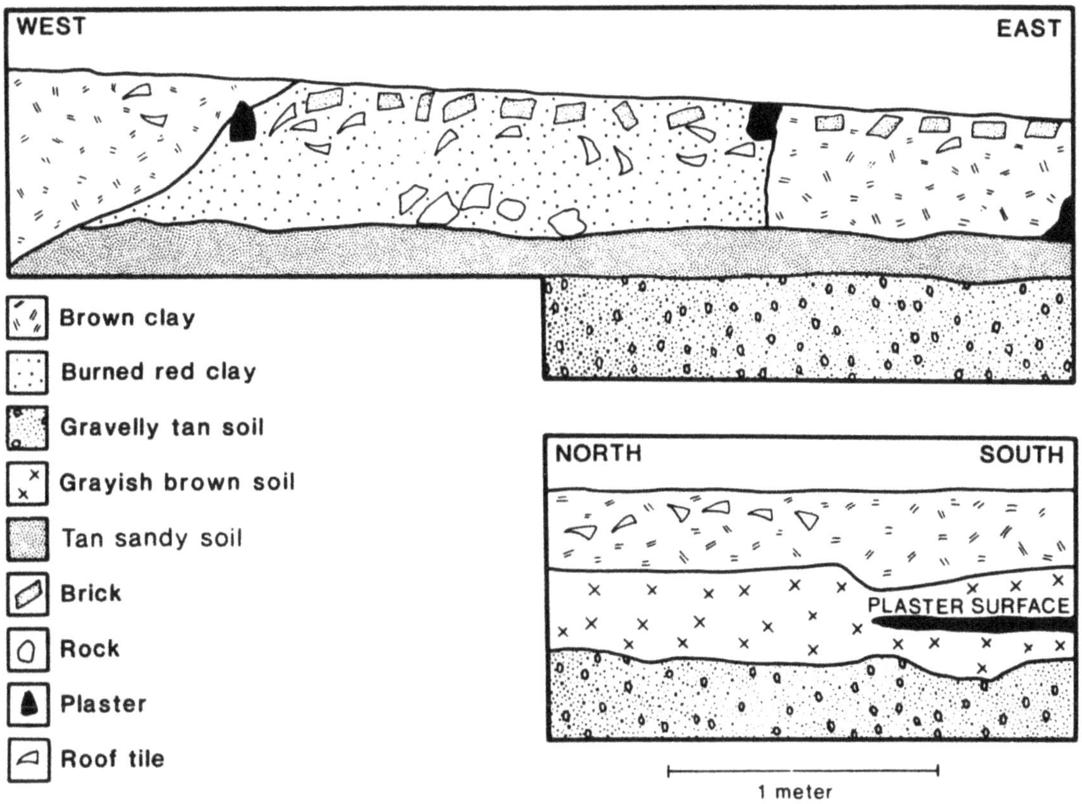


Figure 4.15. North (top) and east (bottom) sections of C1 sounding.

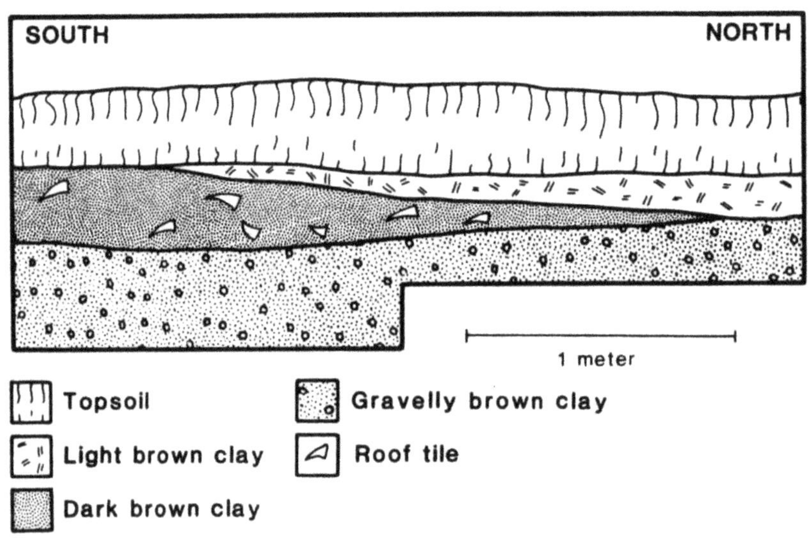


Figure 4.16. West section of D1 sounding.

Below this were 30-40 cm of brown clay that became increasingly gravelly and contained very few artifacts.

H1 (Figure 4.17), a 2-by-5-m sounding located on the valley floor in the eastern part of the site, produced some architectural and human remains. Below 40 cm of topsoil were 50 cm of loose fill that contained a concentration of roof tiles and the remains of a limestone wall (Wall 1). The deposit also contained a moderate amount of pottery but very little bone. Below the fill was a gravelly, red compact clay in which a human skeleton was found nearly intact. The skeleton lay flexed on its right side facing east, and was covered with several flat stones. Below the burial, the soil was devoid of cultural material.

I1, a 2-by-6-m sounding located on the south slope near the crest of the hill, yielded no architecture and a small quantity of pottery toward the bottom of the one-meter-deep unit.

J1, a 2-by-4-m unit situated on the south slope, produced some cultural material but no architecture to a depth of about 50 cm.

## Major Construction/Occupation Phases

Three major construction and/or occupation phases at al-Basra can be discerned on the basis of architectural and sedimentary evidence in the deepest soundings--G1, G2, F3, and F5. These soundings contain a similar sequence of architectural material, characterized by specific types of stone wall construction, and by sedimentary deposits. In some cases, soundings with little or no architectural features can be correlated with this sequence on the basis of ceramic evidence (see Chapter 7).

### Building/Occupation Phase 1

In all likelihood, the earliest building/occupation phase (Phase 1) represents the original settlement of the site by the Arabs in the late eighth or early ninth century. The earliest structures found at the site were built directly on sterile ground, indicating that it had not been inhabited in any substantial way before then. The few walls associated with Phase 1 were built of stone and mortar; only two or three courses of the walls remained intact, possibly because they collapsed or were dismantled by later inhabitants of the city. These walls were covered with the debris of collapsed mud-brick walls and tiled roofs. In one area (G2), the walls were associated with a "flagstone" pavement made of large, flat rocks; the pavement covered a drain which may have been used for water run-off. In addition to residential structures, one of the pottery kilns (F3) in the industrial area may belong to this early phase. The kiln, which was built into sterile ground, was situated stratigraphically _below_ a Phase 2 occupation wall and floor. As we will see in Chapter 5, the pottery found in these early levels are clearly Islamic and are similar in form and paste composition to those present in later levels.

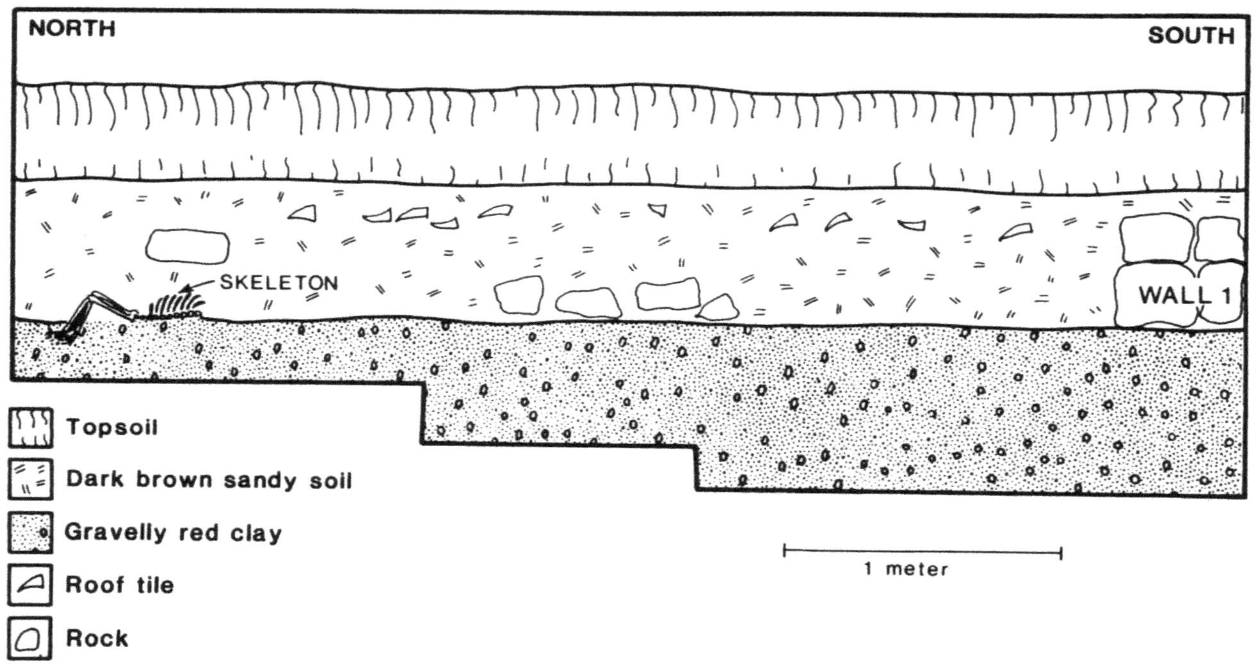


Figure 4.17. East section of H1 sounding.

## Building/Occupation Phase 2

The middle building/occupation phase (Phase 2) represents an extensive reconstruction of the Islamic city. The well-built structures associated with this phase were found across the entire site. They were characterized by well-shaped limestone-and-mortar stone walls, which were typically seven courses high and two courses wide. The walls were built in foundation trenches that were filled with an ashy residue. Some of the walls had very distinctive features, such as a deeply cut channel that ran lengthwise through the center of one wall (G1) and foundation stones that jutted outward, perhaps to support a floor (G2). In many cases, the walls were associated with well-made plaster or cobblestone floors. In contrast to the early walls, many Phase 2 walls survived nearly intact and, in some cases, appeared to have been incorporated into the architectural plan of the following building phase. Many of these structures seem to have survived until the final collapse of the city at the end of Phase 3. Although the stratigraphic evidence is rather inconclusive, it is possible that the F1 pottery kiln and the large cistern (E1) date to the middle occupation phase. The stone and mortar construction of the cistern resembles the construction of Phase 2 stone walls, which are characterized by alternating layers of small, flat stones and large, rectangular rocks. If the cistern were built during the second phase, it is likely that the massive city wall would have also been constructed at the same time. The cistern lies at the foot of the city wall and, conceivably, was built as an integral part of the wall. Although some of these interpretations are tentative (and will remain so until further excavations can be carried out at the site), they summon up a picture of massive construction at the beginning of Phase 2, involving an enormous investment in labor and material and possibly the direction of a central authority.

## Building/Occupation Phase 3

The last major construction/occupation phase (Phase 3) represents a relatively brief period in the history of al-Basra. The stone walls associated with this phase are characterized by their rough appearance, consisting of unshaped rocks mortared with tile, sherd, and bone fragments and resting on a cobble base. They are often found in association with earth or stone floors. In one case, they were adjacent to a large basin-like structure (G2) and, in another location, a stone circle (F3). The structures from this phase appeared to have been incorporated into the architectural layout of the previous occupation. In G2, for example, a stone basin and a pebbled floor were built into a corner of a possible courtyard that was bounded on one side by a Phase 2 limestone and mortar wall and on another side by a Phase 3 wall (see Figure 4.5). In most places, both Phase 2 and Phase 3 walls were covered with the debris of collapsed mud-brick walls and tiled roofs apparently destroyed by fire. In some cases, Phase 3 walls had been dismantled where they protruded above the mud-brick debris.

## Building/Occupation Phase 4

Phase 4 consists of post-urban debris accumulated during the seven centuries or so following the final collapse and abandonment of the city of al-Basra. The presence of a few stone walls <u>above</u> the burned mudbrick deposits shows that the site was occupied intermittently during that time but that no substantial buildings were erected. It is believed that the modern-day village of Jouana-Basra was built around the turn of the twentieth century.

## Spatial Organization of the Urban Center

The architectural evidence also provides some insights into the spatial organization of the Islamic urban center. Although only a small portion of the site has been excavated, the wide distribution of soundings across the hill slopes and on the valley floor allows us to reconstruct, albeit tentatively, the layout of the city.

It is clear that the central portion of the site, situated in the lowest part of the valley between the two hills, was the most intensely settled part of the city throughout its history Excavation units in this area revealed a series of architectural strata and large quantities of cultural material. By contrast, excavation units on the hillsides around the periphery of the site produced relatively little architectural or artifactual remains. While some cultural material may have eroded away, the overall shallowness of soundings in this area (e.g., A1, B1, C1, D1, I1, and J1) indicated that the slopes of the hills were occupied only intermittently, perhaps when the city's population reached its peak. As Table 4.1 shows, pottery and bone density

TABLE 4.1

AL-BASRA POTTERY AND BONE DENSITIES, BY SOUNDING

| Sounding | Location | Sherd Density ($kg/m^3$) | Bone Density ($kg/m^3$) |
|---|---|---|---|
| A1 | north slope | 1.79 | - |
| B1 | north slope | 1.63 | .20 |
| C1 | north slope | .37 | - |
| D1 | north slope | 2.38 | .60 |
| E1 | outside city wall | 1.88 | .59 |
| F1 | industrial area | 5.29 | 3.62 |
| F2 | industrial area | 2.40 | .82 |
| F3 | industrial area | 3.77 | 2.12 |
| F4 | industrial area | 7.67 | 4.71 |
| F5 | industrial area | 7.13 | 3.26 |
| F6 | industrial area | 10.20 | 5.26 |
| G1 | residential area | 5.52 | 1.94 |
| G2 | residential area | 5.97 | 1.16 |
| G3 | residential area | 3.29 | 1.46 |
| H1 | valley floor | 3.13 | 1.13 |
| I1 | south slope | 1.34 | .36 |
| I2 | south slope | 1.86 | .29 |
| J2 | south slope | 7.50 | 1.04 |

in soundings on the valley floor are substantially higher than those on the slopes. The one exception to the pattern, J1, which showed a relatively high density of pottery, would be worth investigating in the future.

In addition, it is fairly clear that the residential and/or commercial quarter was situated toward the center of the site while the industrial sector was located closer to the city walls. As Figure 4.1 shows, residential and/or commercial architecture was discovered primarily in the G soundings, which were concentrated in the middle of the site. In contrast, the industrial sector identified by the pottery kilns found in the F soundings was situated further from the center of the city, although it was still inside the city walls. As Figure 4.18 illustrates, pottery and bone densities in the industrial zone were substantially higher than they were in the residential area, suggesting that the industrial zone may have been the site of dumping activities, perhaps because it was a less densely populated area.

Preliminary though it may be, the spatial layout of al-Basra resembles that found at other medieval Islamic cities in the Maghreb, in particular the city of Fez during the twelfth and thirteenth centuries. According to LeTourneau (1961), the commercial and residential areas of Fez were concentrated in the center of the walled town while gardens and orchards occupied the periphery. Industries, such as potteries, olive oil mills, saw mills, and tanneries, all of which were potential polluters of air and water, were situated near the city gates or along a river that flowed through the city. Small-scale artisans, such as weavers, shoemakers, cobblers, and jewelers were scattered throughout the city's various quarters. LeTourneau's description of Fez, derived primarily from documentary information, provides a detailed picture of a medieval Islamic city established at about same time as al-Basra and thus is summarized at some length here:

> Within the irregular boundary thus defined [by the city wall] there were fairly numerous open spaces near the walls and, from the very first, two vast cemeteries, one for the right bank, to the south of the built-up area, the other for the left bank, to the north. Certainly at that time, as at the beginning of the twentieth century, fairly numerous gardens planted with trees, flowers, and even vegetables, occupied relatively large areas. They were situated, like the cemeteries, along the ramparts, around practically the whole inner circumference of the enciente; at almost no point did the dwellings touch the wall. This was certainly due to breadth of vision on the part of the Almohad builders when marking out the periphery; perhaps it was intended also to allow defenders of the city space to maneuver easily in case of threatened danger. In short, passing the wall did not mean gaining immediate entrance into the city proper, but into a kind of suburb which must have presented a semirural aspect. In that outskirt of the city were established several important industries such as that of the potters, located to the east of the Andalusian Bank, and the olive oil mills, clustered near the gates through which principally olives arrived...[LeTourneau 1961:19-20].

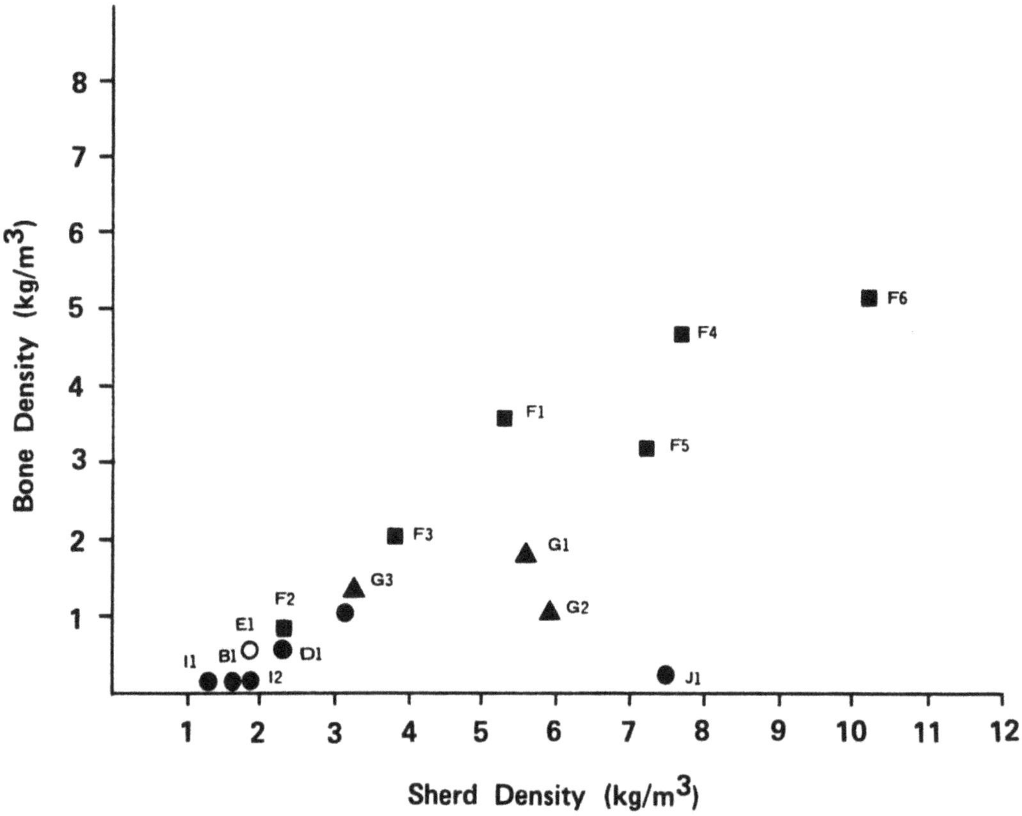


Figure 4.18. Pottery and bone densities in individual soundings at al-Basra. Key: ▲, sounding in residential area; ■, sounding in industrial area; ●, sounding on hill slope; O, sounding outside city wall.

In many respects, the spatial organization of al-Basra seems to resemble that of Fez. The most intensely inhabited area lies in the center of the walled city. The industrial sector is located on the edge of the city. Finally, the most sparsely populated area is situated around the periphery, on the north and south slopes, of the medieval center.

## CHAPTER 5

## DESCRIPTION OF POTTERY, ANIMAL BONE, AND OTHER ARTIFACTS FROM AL-BASRA

In addition to abundant architectural material, the excavations at al-Basra yielded large quantities of ceramic and faunal material, as well as small amounts of metal, glass and stone objects. Because this material, in particular the pottery, is used in the development of a relative chronology for the site and a study of craft specialization, it is described in some detail in this chapter. The ceramic material is discussed in terms of its classification by ware, form and decorative technique. The faunal and other artifactual remains are described briefly at the end of the chapter.

### Ceramic Material

More than 100,000 potsherds were recoverd from the site of al-Basra during the 1980 and 1981 field seasons. All of this material was collected and weighed by provenience unit (i.e., level and locus) and sent to the project's laboratory in Tangier for processing and storage. During both field seasons, a small part of the material was examined and a preliminary ceramic typology was set up (Redman 1983, in press; Redman et al. 1980).

In 1983, I reclassified a large portion of the al-Basra collection --including nearly 75,000 body and diagnostic sherds from 142 provenience units (Table 5.1)--according to a revised and expanded typology. The sherds were recounted and reweighed by provenience unit and certain attributes, such as bloating and smudging, were noted. In addition, the nearly 12,000 diagnostic sherds, i.e., rims, bases, and handles

TABLE 5.1

COUNTS AND WEIGHTS OF AL-BASRA POTTERY (142 PROVENIENCE UNITS)

| Sherd Type | Sherd Count (No.) | Sherd Count (%) | Sherd Weight (Kg) | Sherd Weight (%) |
|---|---|---|---|---|
| Body Sherds | 62,651 | 83.5 | 677 | 70.0 |
| Diagnostic Sherds: | | | | |
| Rims and Others | 6,016 | 8.0 | 105 | 11.0 |
| Bases | 4,653 | 6.2 | 137 | 14.0 |
| Handles | 1,671 | 2.2 | 45 | 5.0 |
| Total | 12,340 | 16.5 | 287 | 30.0 |
| Grand Total | 74,991 | 100.0 | 963 | 100.0 |

from 133 provenience units, were coded by ware and form category. In general, pottery wares were delineated on the basis of paste composition, such as color, texture, and type and size of inclusions, and forming, firing, and finishing attributes. Vessel forms were usually identified by the size and shape of the rim and neck and, when available, the body and base.

## Ware Categories

The collection consists of seven basic ware categories--handmade, sandy buff, fine buff, very fine buff, cream-colored, coarse, and glazed wares. In some cases, these groups overlap somewhat with one another; in others, they encompass a variety of different wares. In general, however, each ware group represents the use of a distinctly different raw material and/or manufacturing technology.

### Handmade wares

A small proportion (5.4%) of the al-Basra pottery consists of handmade wares (Table 5.2). These vessels were formed by hand rather than thrown on a wheel. On the whole, they were made by rolling the clay into coils, placing the coils on top of each other, and smoothing the walls with a flat tool to finish the vessel. The vessels were fired in the open air under relatively low temperatures rather than in a kiln. This ware seems to be found almost exclusively in the form of cooking vessels.

Handmade wares are characterized by a paste that contains large inclusions, such as grog (crushed pottery), grit, limestone, and vegetal matter. The paste varies in color from pale buff to very dark gray and is relatively soft and porous. The vessel walls are thick, often measuring greater than 1 cm. The exterior surfaces are frequently blackened by carbon, most likely as a result of reducing gases from a cooking fire. Thick walls and large inclusions, such as grog, have been shown to allow vessels to withstand repeated heating over a fire and thus to prevent breakage induced by thermal shock (Steponaitis 1983). The al-Basra handmade vessels are commonly decorated with a finger- or tool-impressed strip of clay applied to the exterior of the vessel below the rim (Figure 5.1b, c).

In general, handmade wares are believed to have been produced by domestic potters, i.e., non-specialist potters, generally women, who made vessels for their own use, very much as they do today in many parts of North Africa (Balfet 1965). In fact, several women in Jouana-Basra, the small village located on the site, produce handmade cooking vessels for their own domestic use; instead of coiling the clay, however, they use a slab technique to build their vessels.

Handmade pottery is seldom mentioned in the archeological literature for North Africa, although it was inevitably present during Islamic times. In a rare reference to it, Bel (1914) noted that handmade sherds, characterized by thick walls, poor kneading, and open-air firing, were present in deposits associated with a tenth- to eleventh-century kiln in the city of Tlemcen (formerly Agadir) in Algeria. In order to explain their association with wheelmade pottery in kiln-

TABLE 5.2

COUNTS AND PERCENTAGES OF AL-BASRA WARES,
BY DIAGNOSTIC PART
(133 PROVENIENCES)

| Diagnostic Part | Total (No.) | Handmade (No.) | Handmade (%) | Sandy (No.) | Sandy (%) | Fine (No.) | Fine (%) | Very Fine (No.) | Very Fine (%) | Cream-Colored (No.) | Cream-Colored (%) | Coarse (No.) | Coarse (%) | Glazed (No.) | Glazed (%) |
|---|---|---|---|---|---|---|---|---|---|---|---|---|---|---|---|
| Rims | 5,387 | 465 | 8.6 | 568 | 10.5 | 2,762 | 51.3 | 779 | 14.5 | 648 | 12.0 | 95 | 1.8 | 71 | 1.3 |
| Bases | 4,312 | 114 | 2.6 | 574 | 13.3 | 2,605 | 60.4 | 524 | 12.2 | 364 | 8.4 | 89 | 2.0 | 42 | 1.0 |
| Handles | 1,473 | 6 | 0.4 | 158 | 10.7 | 790 | 53.6 | 170 | 11.5 | 226 | 15.3 | 45 | 3.1 | 78 | 5.3 |
| Other | 233 | 30 | 12.9 | 10 | 4.3 | 100 | 42.9 | 18 | 7.7 | 40 | 17.2 | 25 | 10.7 | 10 | 4.3 |
| Total | 11,405 | 615 | 5.4 | 1,310 | 11.5 | 6,257 | 54.9 | 1,491 | 13.1 | 1,278 | 11.2 | 254 | 2.2 | 201 | 1.8 |

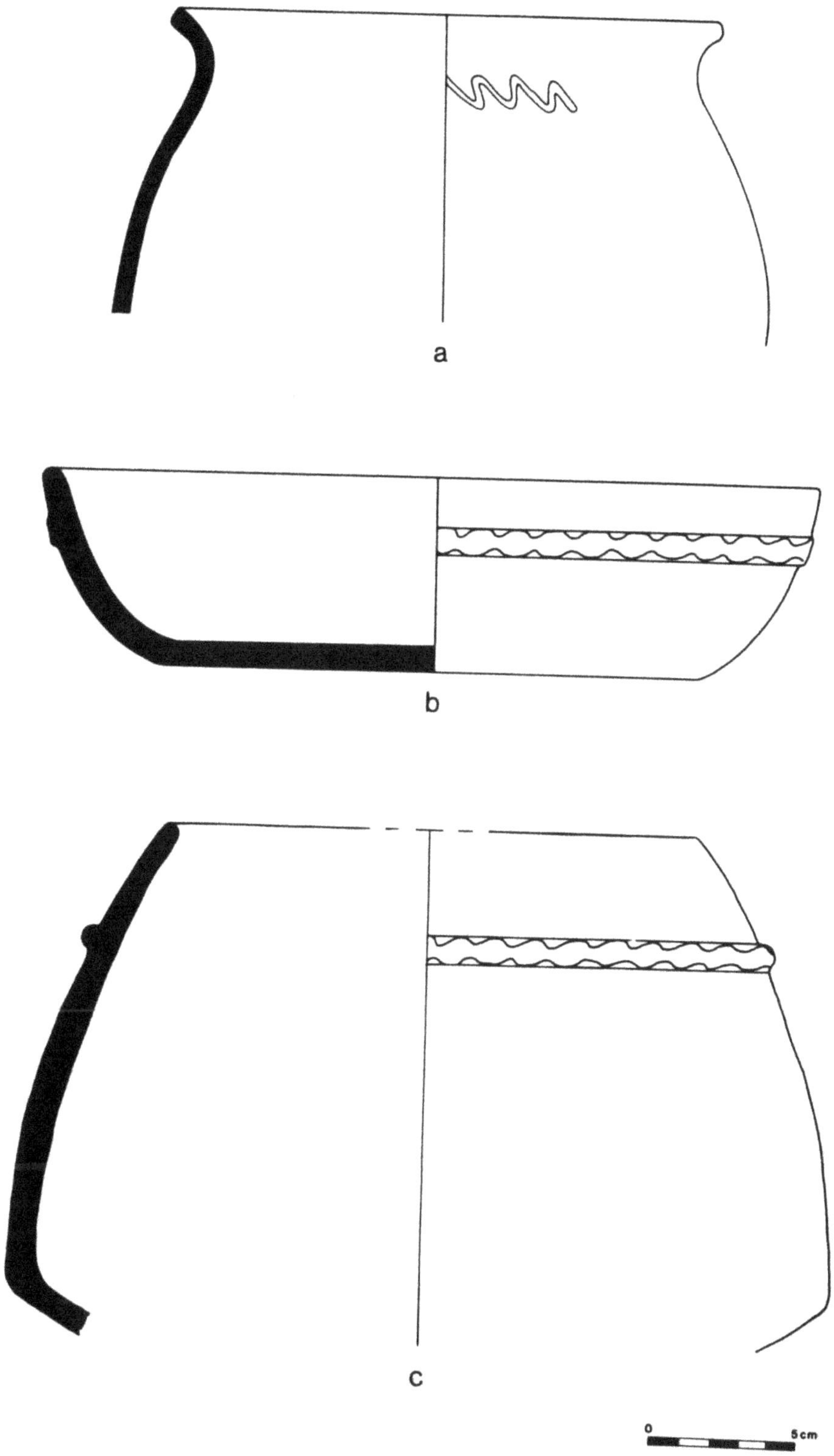


Figure 5.1. Handmade and coarse cooking vessels. (a) Coarse ware jar with flattened, everted lip and low neck decorated with incised zigzag motif (Form 54). (b) Shallow handmade cooking vessel with finger-impressed strip below rim (Form 10). (c) Deep handmade cooking vessel with finger-impressed strip below rim (Form 11).

related deposits, he suggested that the handmade material was made by the potters' wives for home use.

More recently, handmade wares have been reported in the finds for a number of Islamic sites in northern Morocco, including Ksar es-Seghir, Nakur, and Badis, which were excavated by Redman. In order to explain the variation in the proportion of handmade sherds from site to site, Redman (1983) has suggested that the relatively low amounts found at cities like al-Basra and Ksar es-Seghir indicates that these centers had weak ties with rural pottery producers, under the assumption that handmade wares were made in the countryside. On the other hand, the relatively high proportions at Nakur and Badis suggests that these cities maintained strong ties to rural producers. Other explanations are equally plausible. If handmade pottery was produced at or near the urban center, as chemical analyses of the al-Basra material suggest (Benco n.d.a), the low proportion may refect the inhabitants' preference for and accessibility to other kinds of cooking vessels, such as copper pots, which were commonly used in the Islamic world and are known historically to have been imported into Morocco from Spain during this time (Goitein 1973:17, 51-56). Unlike Nakur and Badis, al-Basra and Ksar es-Seghir were both located along one of the major trade routes connecting sub-Saharan Africa with the Mediterranean, which would have given them access to material moving from one to the other.

Wheelmade wares

By far, the largest part (95%) of the al-Basra collection consists of pottery that was thrown on a wheel and fired in a kiln. Unlike the hand-built wares, the wheelmade plainwares were very likely produced by artisans in pottery workshops and sold in urban and rural markets. With the exception of glazed wares, most of the wheelmade pottery found at al-Basra seems to have been made at the site (Benco n.d.a).

<u>Buff wares--sandy, fine, and very fine.</u> The most common wheelmade pottery consists of a buff ware that is usually undecorated, except for some occasional incising, slipping or painting. As a group, buff wares are characterized by a paste that ranges in color from buff to gray on the surface and is often gray in the core. The variation in surface color is believed to have been caused by an uneven oxidizing or reducing atmosphere in the kiln during firing, rather than to any deliberate attempt by potters to achieve a particular hue. As a result, both buff and gray sherds are lumped together as buff wares for the purposes of this study. Buff wares have relatively hard surfaces and fairly thin walls. The paste contains small amounts of calcite, feldspars, mica, and iron oxides and varying amounts of quartz. The variation in the size and quantity of quartz forms the basis for dividing buff wares into three groups: sandy, fine, and very fine. As we will see later, the variation represent a shift in the manufacturing technology employed by the potters at al-Basra.

Sandy wares, which constitute about 12% of all al-Basra pottery (Table 5.2), represent the smallest of the three buff-firing groups. This ware is characterized by a paste that contains relatively large, densely-packed quartz particles, giving it a gritty texture on the surface. Because of the coarseness of the paste, the color separation between buff surface and gray core is usually blurred. About 15% of sandy sherds are covered with a white slip on the exterior surface and none are painted.

Fine wares represent the largest ware group at al-Basra, constituting 55% of the entire collection (Table 5.2). Although they may overlap somewhat with sandy and very fine wares, fine wares form a fairly distinct and homogeneous group. They are characterized by a paste that contains a moderate amount of quartz particles--resembling a loose network of grains under a 10x lens--and a relatively smooth surface. The color separation between surface and core is fairly distinct because of the fineness of the paste. Less than 10% of these wares are slipped and less than 1% are painted.

Very fine wares amount to about 13% of the al-Basra collection (Table 5.2). This ware is characterized by a paste that contains very small quartz inclusions and consequently a smooth, almost polished-looking, surface. Because of the fineness of the paste, the color separation between surface and core is extremely sharp, giving it a "sandwiched" appearance (Redman 1983, n.d.; Redman et al. 1980). Of all the buff wares, very fine wares have the hardest and thinnest walls. About 11% of these wares are slipped and less than 1% are painted.

Cream-colored wares. These wheelmade wares make up about 11% of the al-Basra pottery (Table 5.2). They are distinguished by a highly calcareous cream-colored paste that ranges in hue from light brown to light olive-green, very likely the result of variable atmospheric conditions in the kiln. The paste is relatively soft and porous and has a powdery surface texture; it contains calcite, quartz, and iron oxide inclusions. Of all the wares found at al-Basra, cream-colored sherds are most frequently painted. About 13% (81 of 648) of the rims are decorated with a simple linear and dot motif which may be epigraphic in nature (Figure 5.2). The iron oxide paint ranges in color from light red to dark brown, the result of variable atmospheric conditions in the kiln.

Painted cream-colored wares, which are often found in the form of a spoutless pitcher with a handle, resemble a type of pottery described as "Berber ware" and found at other Islamic sites in Spain (Velaquez-Bosco 1912, Zozoya 1969) and in the Maghreb (Bel 1914). This pottery has been called Berber ware because of its similarity in form and decoration to pottery produced by contemporary Berbers in the Rif mountains (van Gennep 1911). It differs, however, in technique, since the archeological pottery is wheel thrown and kiln fired and the modern-day ware is usually handmade and fired in an open-air pit.

Coarse wares. This group encompasses a variety of different coarse wares, which together constitute about 2% of the al-Basra pottery (Table 5.2). In general, this group is characterized by a coarse paste and/or thick vessel walls. One variety has a pinkish-buff colored paste with large calcite inclusions and is usually found in the form of a large storage jar. Another has a finely-textured, brick red paste and is frequently found in the form of heavy jars and basins, such as those recoverd from the city water cistern at al-Basra. According to

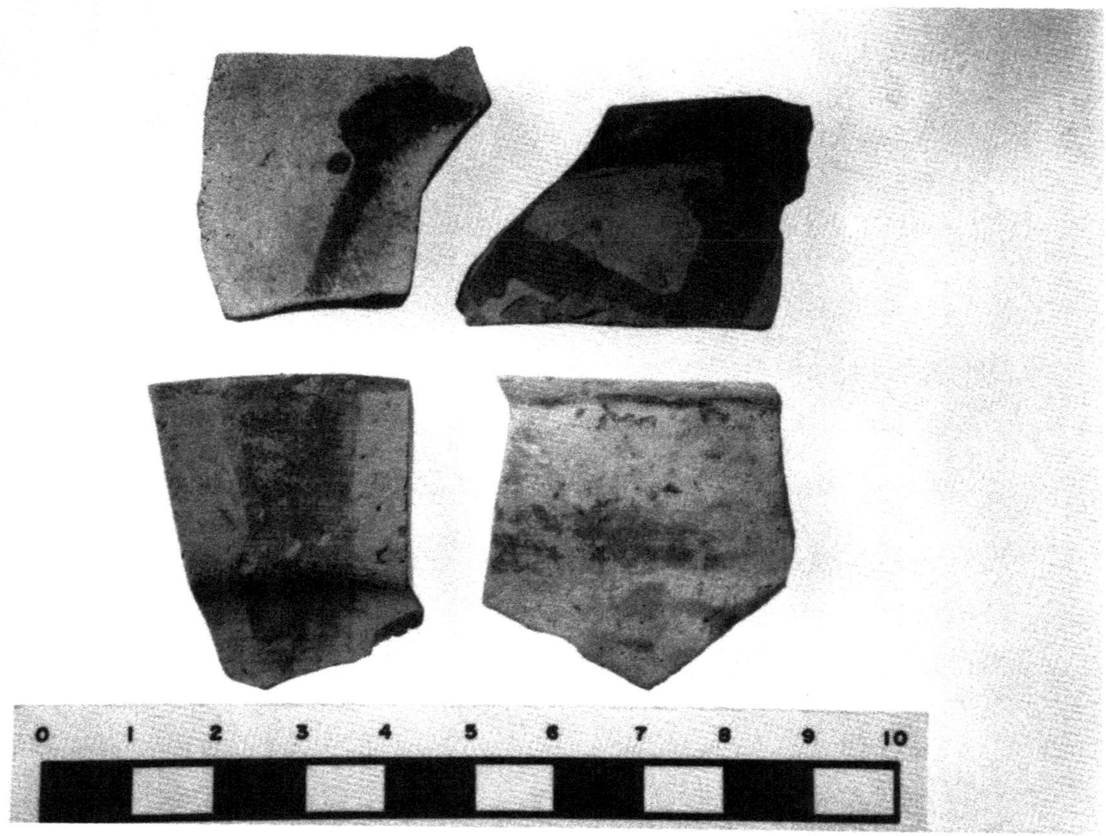


Figure 5.2 Red-painted cream-colored wares.

chemical studies, both wares appear to have been made at al-Basra (Benco n.d.a).

Glazed wares. This small group of wheelmade sherds, which represent about 2% of the al-Basra collection (Table 5.2), consists of glazed wares. Despite their small number, this group encompasses a wide variety of decorated and undecorated monochrome and polychrome types (Table 5.3). They also differ in paste composition. Unlike the other wares described above, most of the glazed pottery seems to have been made at other production centers in North Africa or Spain and brought to al-Basra through various commercial or social ties.

TABLE 5.3

COUNTS AND PERCENTAGES OF AL-BASRA GLAZED POTTERY

| Glaze Type | Count (No.) | Percentage (%) |
|---|---|---|
| Undecorated Glaze: | | |
| Green | 117 | 34.2 |
| Honey | 56 | 16.4 |
| Honey and Green | 17 | 5.0 |
| Brown | 34 | 9.9 |
| Yellow | 6 | 1.8 |
| Light Green | 6 | 1.8 |
| Honey on White Slip | 22 | 6.4 |
| Green on White Slip | 11 | 3.2 |
| Honey on Green Paint | 19 | 5.6 |
| Decorated Glaze: | | |
| Manganese-on-Honey | 25 | 7.3 |
| Green/Manganese | 3 | 0.9 |
| Glaze Painted | 15 | 4.4 |
| *Cuerda Seca* | 2 | 0.6 |
| Incised/Punctated under Glaze | 9 | 2.6 |
| Total[1] | 342 | 100.0 |

[1] The glazed pottery total includes all rim, base, handle, and body sherds recovered at the site.

The most common type is a green- or honey-colored, lead-based glazed ware with no decoration. This ware is usually glazed on both exterior and interior surfaces although, in some cases, it is glazed on the exterior only. A few examples have a green-colored glaze on one side and a honey-colored glaze on the other, probably reflecting the range of colors that lead-based glazes can acquire, depending on atmospheric conditions in a kiln. Lead-based glazes become green in an oxidizing atmosphere and honey colored in a reducing atmosphere (Rye 1981, Rhodes 1981).

A smaller group consists of undecorated brown (manganese oxide base), yellow, and light green glazed wares, and honey- or green-colored glazed wares over a white slip or green paint.

A small proportion of the glazed pottery is decorated. The most common type is a manganese-painted, honey-colored glazed ware with a dark brown to purplish-brown manganese design appears on a clear, honey-colored glaze (Figure 5.3a). The design usually consists of a simple linear or curvilinear motif painted on the interior surface of the vessel. Many of these vessels are flat-bottomed, shallow bowls. A rarer type has a light yellow glaze over a green-painted motif; in one example, the motif consists of a floral pattern with a dark green curving stem and light green buds (Figure 5.3d), somewhat similar to the design on a plate found at the early Islamic site of 'Abbasiyya in Tunisia (Marcais 1928:Figure 16).

A less common type is what I call a glaze painted ware, which is similar to the _ceramique à réservé_ described by Marcais (1928). Some examples are decorated with a green or honey-colored glaze that has been dribbled or painted on the surface with a brush without any apparent guidelines. Others have motifs that have been outlined in a brownish-black paint first and then filled in with a glaze. The designs are painted directly onto the surface and cover only a small portion of the entire vessel (Figure 5.3e, f). Their pastes are generally cream colored and relatively friable. An important exception is a green glaze painted base fragment (Figure 5.3f) made of a fine buff-firing paste that is very similar in chemical composition to other local al-Basra wares (Benco n.d.a). It is possible that this sherd represents an experiment by an al-Basra potter to imitate the glaze-painted wares that were imported into the site.

A few sherds are decorated in _cuerda seca_ style, in which a design is traced with a non-fluxing substance, like paint, and the entire vessel surface is covered with variously colored glazes; the paint prevents the glazes from running into each other during firing (Figure 5.3c).

Finally, a few glazed fragments are decorated with incised or punctated designs under a green- or honey-colored glaze. The designs consist of simple lines etched into the soft clay and later covered with glaze. This type of decoration seems to appear most frequently on lamp bases.

## Form Categories

The al-Basra collection contains nearly 40 different vessel forms (Table 5.4), but only 11 of these are commonly found at the site. The 11 forms, which include small and large storage jars, pitchers, cooking vessels, bowls, and oil lamps, together constitute about 95% of all the rim sherds. The remaining 29 forms occur in very small quantities (from one to 40 sherds apiece) and make up about 5% of the collection.

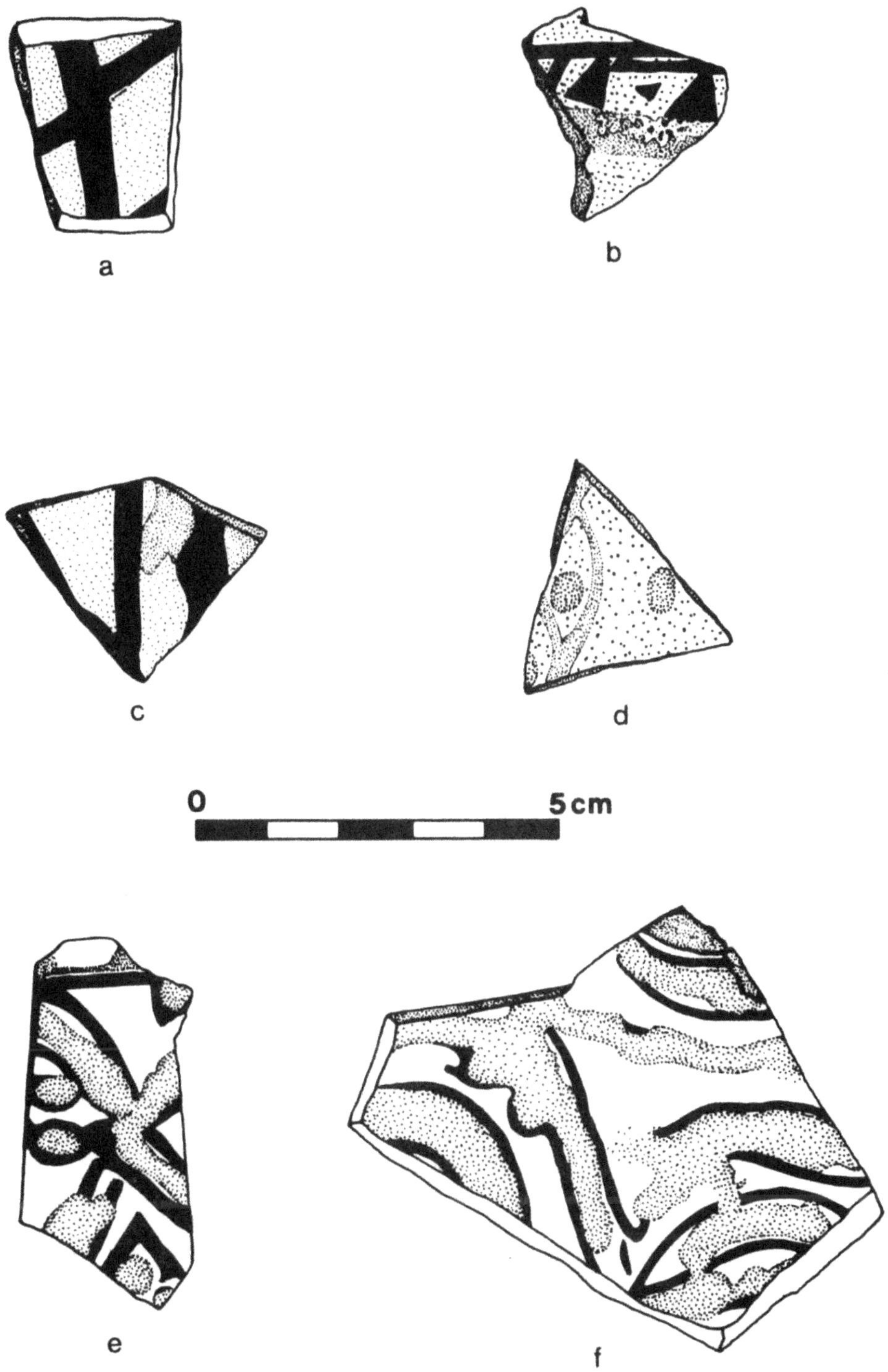


Figure 5.3. Decorated glazed wares. (a) Manganese-painted, honey-colored glazed body sherd, with linear motif. (b) Manganese-on-light-yellow (bisque) glazed body sherd, with geometric motif. (c) cuerda seca body sherd, with green glaze separated by dark brown painted lines. (d) Green-on-light-yellow (bisque) glazed body sherd. (e) Glaze-painted rim sherd with green glaze floral (?) motif outlined in dark brown paint. (f) Glaze-painted base sherd, with green glaze motif outlined in dark brown paint, possibly of local al-Basra production.

TABLE 5.4

COUNTS AND PERCENTAGES OF AL-BASRA FORMS
(133 PROVENIENCES)

| Functional Category: Form Number | Count (No.) | Percentage (%) |
|---|---|---|
| Cooking Vessels: | | |
| 10 | 339 | 6.3 |
| 11 | 95 | 1.8 |
| 25 | 20 | 0.4 |
| | --- | --- |
| Total | 454 | 8.5 |
| Bowls: | | |
| 12 | 84 | 1.6 |
| 13 | 34 | 0.6 |
| 14 | 2 | 0.0 |
| 15 | 4 | 0.0 |
| 16 | 6 | 0.0 |
| 17 | 3 | 0.0 |
| | --- | --- |
| Total | 133 | 2.5 |
| Basins: | | |
| 19 | 14 | 0.3 |
| 20 | 5 | 0.0 |
| | --- | --- |
| Total | 19 | 0.3 |
| Jars: | | |
| 40 | 872 | 16.2 |
| 41 | 925 | 17.2 |
| 42 | 137 | 2.5 |
| 43 | 815 | 15.1 |
| 44 | 226 | 4.2 |
| 49 | 12 | 0.2 |
| 50 | 10 | 0.2 |
| 52 | 20 | 0.4 |
| 53 | 27 | 0.5 |
| 54 | 24 | 0.5 |
| 55 | 1 | 0.0 |
| 72 | 33 | 0.6 |
| 74 | 106 | 2.0 |
| 76 | 1 | 0.0 |
| 78 | 5 | 0.0 |
| | ----- | ---- |
| Total | 3,214 | 59.7 |

TABLE 5.4--Continued

| Functional Category: Form Number | Count (No.) | Percentage (%) |
|---|---|---|
| **Pitcher:** | | |
| 48 | 1,076 | 20.0 |
| Total | 1,076 | 20.0 |
| **Beaker:** | | |
| 56 | 9 | 0.2 |
| Total | 9 | 0.2 |
| **Flasks:** | | |
| 68 | 25 | 0.5 |
| 69 | 7 | 0.1 |
| 70 | 4 | 0.0 |
| 71 | 1 | 0.0 |
| 79 | 2 | 0.0 |
| Total | 39 | 0.7 |
| **Lamps:** | | |
| 93 | 417 | 7.7 |
| 94 | 4 | 0.0 |
| Total | 421 | 7.8 |
| **Cup:** | | |
| 110 | 23 | 0.4 |
| Total | 23 | 0.4 |
| Grand Total | 5,388 | 100.1 |

A variety of bases, handles, and miscellaneous ceramic objects are also present in the collection and are described briefly at the end of this section.

Small and large storage jars

The most common vessel form found at al-Basra is a small storage jar (Forms 41, 42, 43). These jars are characterized by short, curved necks, pear-shaped bodies, and flat bases (Figures 5.4 and 5.5). They are about 18 cm in height. The shoulders are sometimes incised with horizontal lines and the lower part of the bodies are often ribbed. These jars are almost always made of a buff-firing paste.

Small storage jars can be divided into three distinct forms on the basis of rim shape. Form 41, which occurs most frequently (17%), is characterized by a triangular lip, in which the rim is partially folded over on the exterior to form a triangular cross-section (Figure 5.4b). Form 42, which is much less common (3%), is distinguished by a banded rim, in which the rim is pressed over completely on the exterior to form a flat band (Figure 5.4a). Form 43, which occurs almost as frequently as Form 41 (15%), is characterized by a banded rim with a groove made by pressing a rounded tool or finger in the soft clay as the vessel rotated on a wheel (Figure 5.5a-d).

Two large forms of storage jars also occur at al-Basra, although much less frequently (2-4%). Form 74 is similar in shape to Form 41, but has larger dimensions (Figure 5.6). Its rim diameter is greater than 18 cm, its height is about 40 cm, and its walls are relatively thick. Otherwise, it is characterized by the same triangular rim, short curved neck, and pear-shaped body as Form 41. Its base is slightly convex rather than flat. Its neck and shoulder are frequently decorated with bands of horizontal or wavy incised lines. The jar is usually made of a buff ware. The width of the mouth suggests that this jar may have been used to store dry goods, such as grains, rather than liquids.

Form 44, on the other hand, is characterized by a tall, narrow neck that often has a ridge on the exterior at about midpoint, possibly to tie a cord around (Figure 5.7a-b). Its rim is folded over to form a triangular or snub-nosed profile. The neck curves into the shoulder gradually. The sloping shoulder is frequently decorated with comb incising. The body is probably round and slightly elongated and the base relatively flat, although the lower part of this vessel has not been reconstructed. The jar is generally made of a buff ware. The height and narrowness of the neck suggest that this jar may have been used for storing liquids, such as water or oil.

Storage jars that resemble the ones found at al-Basra are seldom mentioned in the literature on Islamic pottery. Among the few possible analogues that have been illustrated are several rim fragments with triangular rims and short, curved necks recovered by Bel at Tlemcen (1914:Figures 15, 17, 18) and a couple of rims from Nakur (Redman 1983: Figure 8). Other examples may exist but have not been published. A number of storage jars known from eleventh- or twelfth-century contexts differ in shape from the al-Basra specimens. Jars from the Hammadid sites of Qal'a des Banu Hammad and Bougie in Algeria, for example, feature high, narrow necks with elaborately painted and etched decorations (Golvin 1965:Figures 78, 79, 82; Marcais 1913:Plate 4, 1916:Plate 23); those from the Almohad fortress of Dchîra in Morocco have inverted, conical bodies (Terrasse 1937:Plate II); those from Salé in Morocco feature short, vertical necks and flat, everted rims (Delphy 1955: Plates 4, 7).

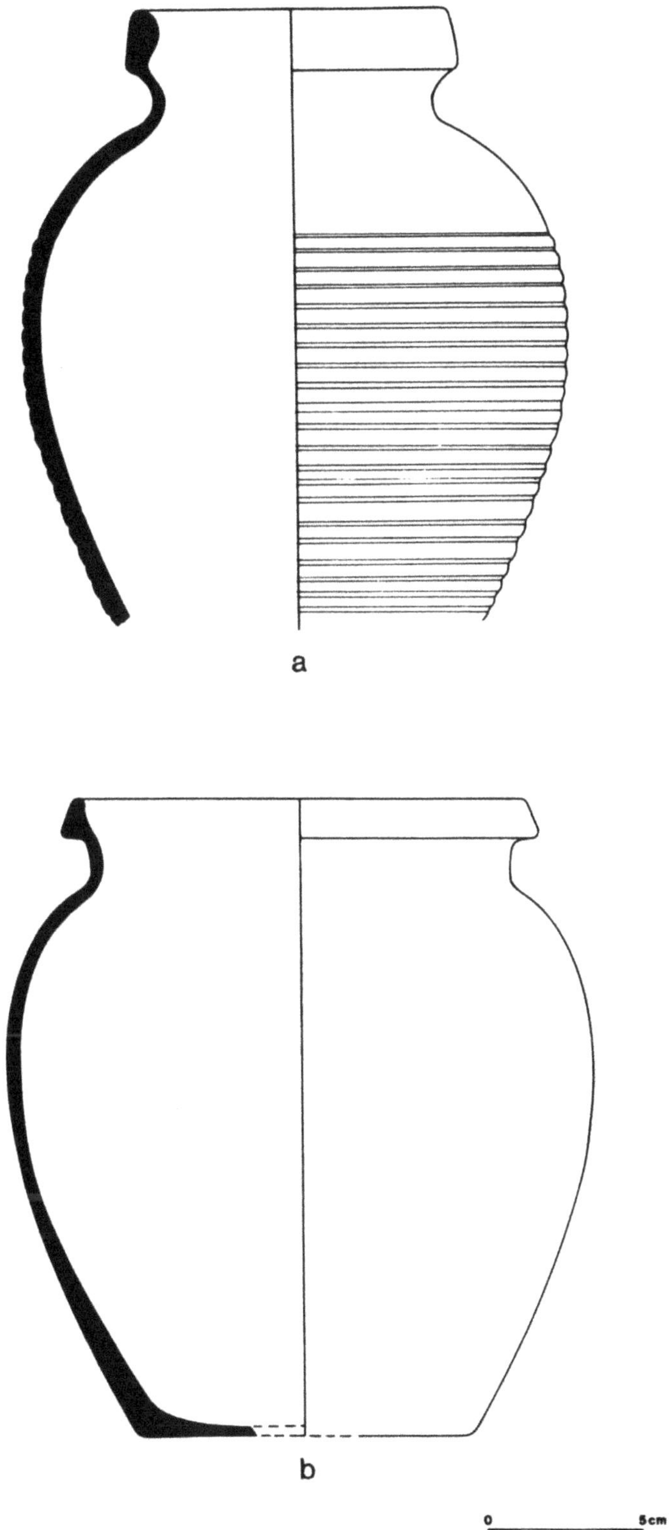


Figure 5.4. Small storage jars. (a) Small storage jar with band rim and ribbed body (Form 42). (b) Small storage jar with triangular rim (Form 41).

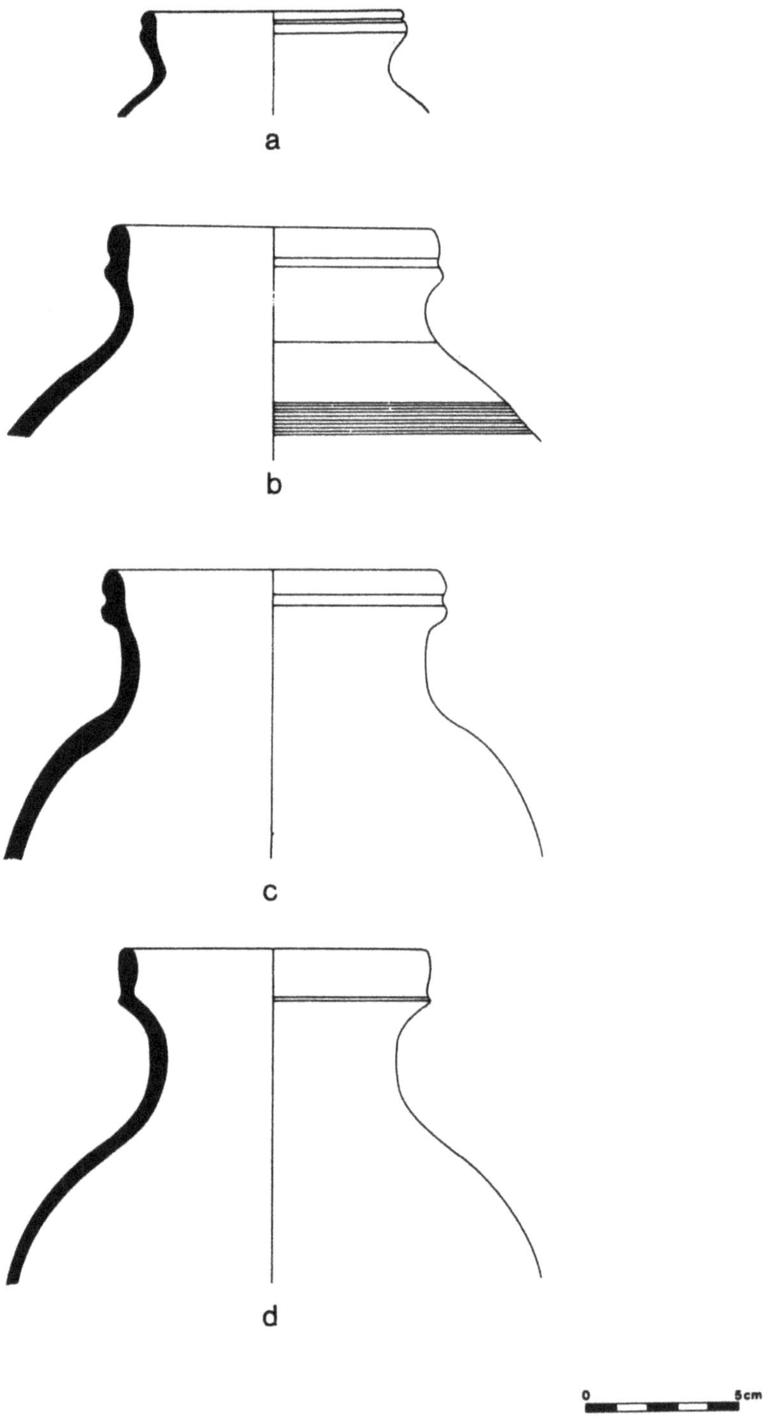


Figure 5.5. Small storage jars with grooved-band rims (Form 43).

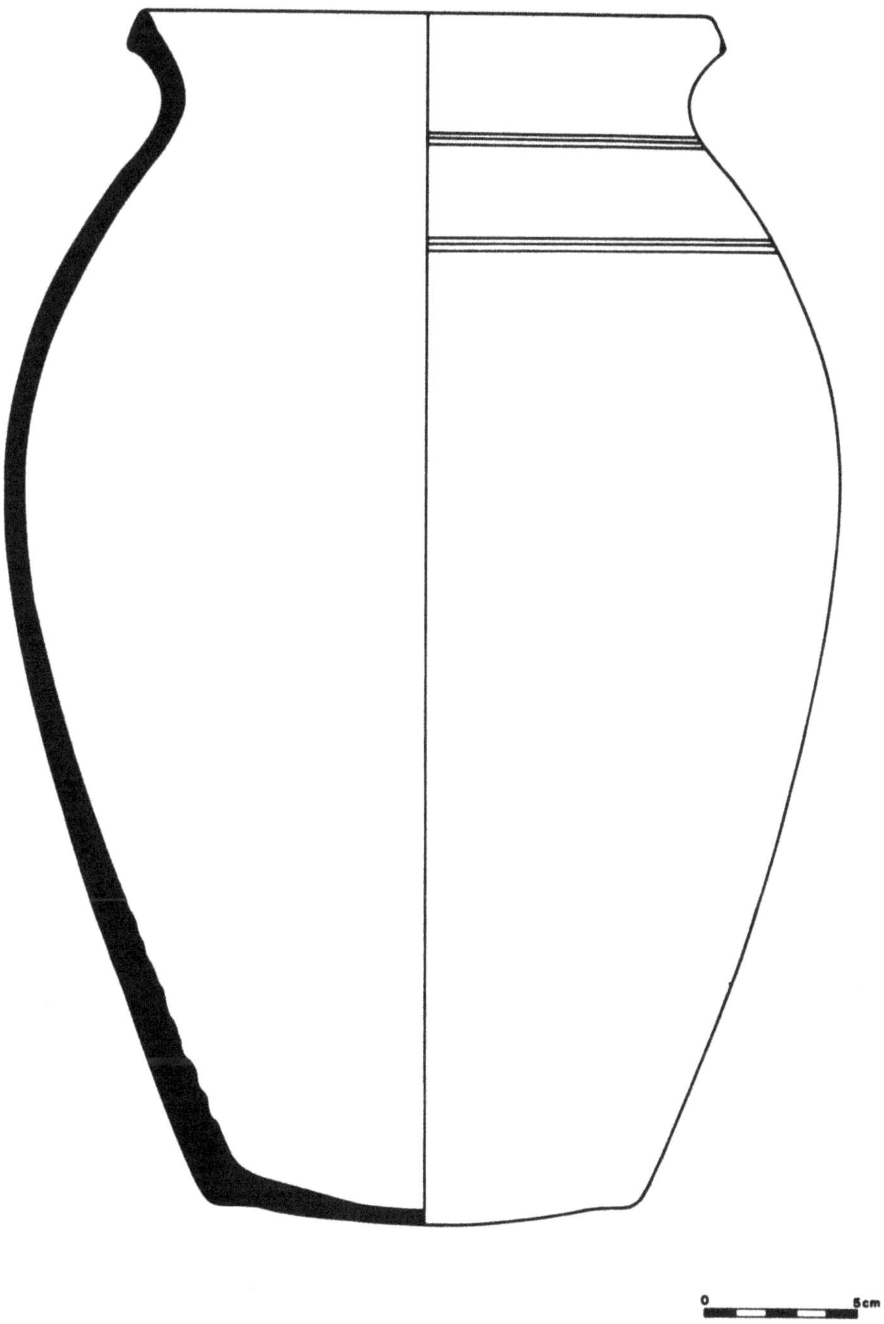


Figure 5.6. Large storage jar with triangular rim (Form 74).

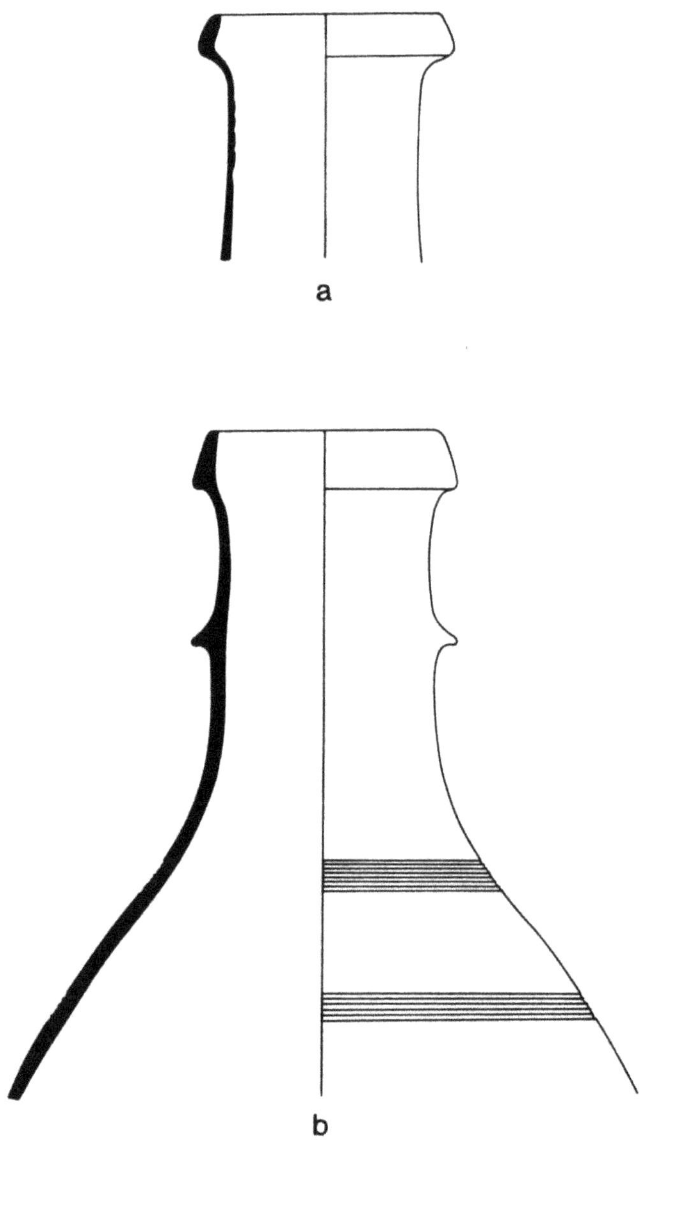


Figure 5.7. Tall-necked water storage jar (Form 44).

### Pitchers

Another form that occurs frequently at al-Basra (20%) is a spoutless pitcher, often with a handle (Figure 5.8a-d). The vessel is characterized by a simple rounded or slightly beveled lip, a tall, nearly vertical neck, and a rounded or sharply pointed carina at the shoulder. The lower body has straight sides that slope inward toward a flat base. The handle, which is attached directly to the rim or slightly below it, extends to just below the shoulder. This vessel is usually made of a cream-colored or buff ware. As noted earlier, cream-colored pitchers are sometimes decorated on the neck and shoulder with red- or brown-painted motifs (Figure 5.2).

Described as Berber ware because of its similarity to ethnographically known examples from the Rif mountains, this type of pitcher occurs at other Islamic sites, among them the tenth- or eleventh-century site of Tlemcen in Algeria (Bel 1914:Figure 40) and the tenth-century Umayyad palaces at Medina az-Zahra and Medina Elvira in southern Spain (Zozoya 1980a:Figures 2, 3, 4; Velazquez Bosco 1912:Plate 39).

### Carinated jars

Another jar form that occurs fairly frequently at al-Basra (16%) is a wide-mouthed, squat jar with a carinated shoulder (Form 40). This form features an everted rim and a body that slopes outward to a sharp, often ridged or grooved carina; below the carina, the body slopes inward toward a flat base (Figure 5.9a-d). The lower part of the body is often ribbed or grooved. The jar is generally made of a buff ware. Other examples of this form of jar have not been found in other publications on early Islamic pottery from Spain or North Africa.

### Shallow bowls

A less common form (2%) in the al-Basra collection is a shallow bowl with a flat base (Form 12). The bowl is usually characterized by a simple, rounded lip and smoothly curving sides (Figure 5.10b,c). It is made predominantly of a buff or cream-colored ware. A few examples with more vertical sides are glazed (Figure 5.10a). A couple of other variants include Form 13, which has a bevelled lip (Figure 5.10d), and Form 14, which has an everted lip.

### Nosed lamps

A relatively common form (7.7%) is a nosed oil lamp (Form 93). The lamp has a flattened, round bowl with a short, curved neck and flaring rim (Figures 5.11a, 5.12). A small handle is attached to the rim and extends to the bowl. A long nose is attached to the opposite side of the bowl. The nose is frequently charred at the tip where the flame burned. The lamp is usually made of a buff- or cream-colored paste. This type of unglazed oil lamp occurs frequently at other Islamic sites in North Africa (Bel 1914:Figure 20; Delphy 1955:Plate 9; Golvin 1965: Figure 76) and Spain (Zozoya 1980a:Figure 11; Velaquez Bosco 1912:Plate 39).

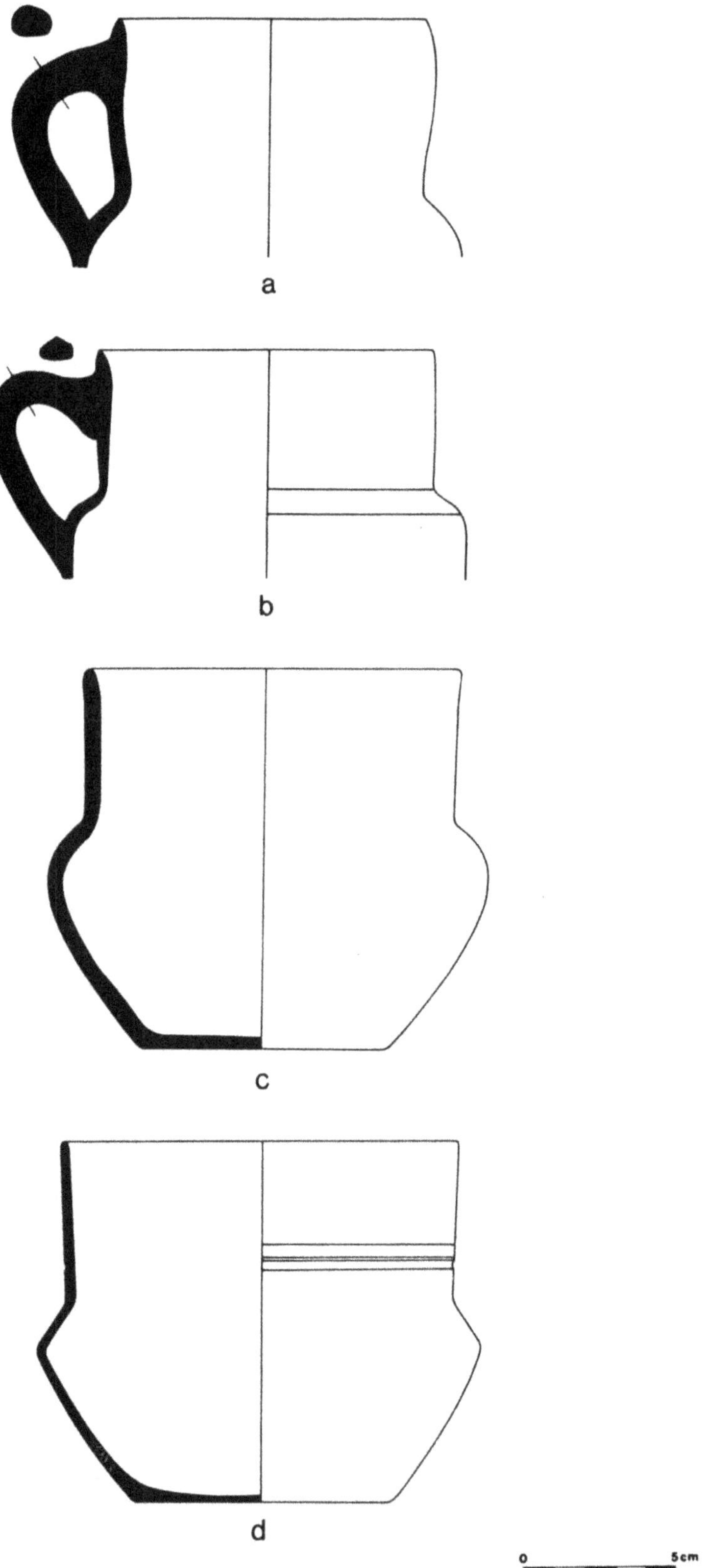


Figure 5.8. Spoutless pitchers (Form 48). (a) Pitcher with handle and rounded shoulder. (b) Pitcher with handle and sharply carinated shoulder. (c) Pitcher with rounded shoulder. (d) Pitcher with sharply carinated shoulder and incised lines on neck.

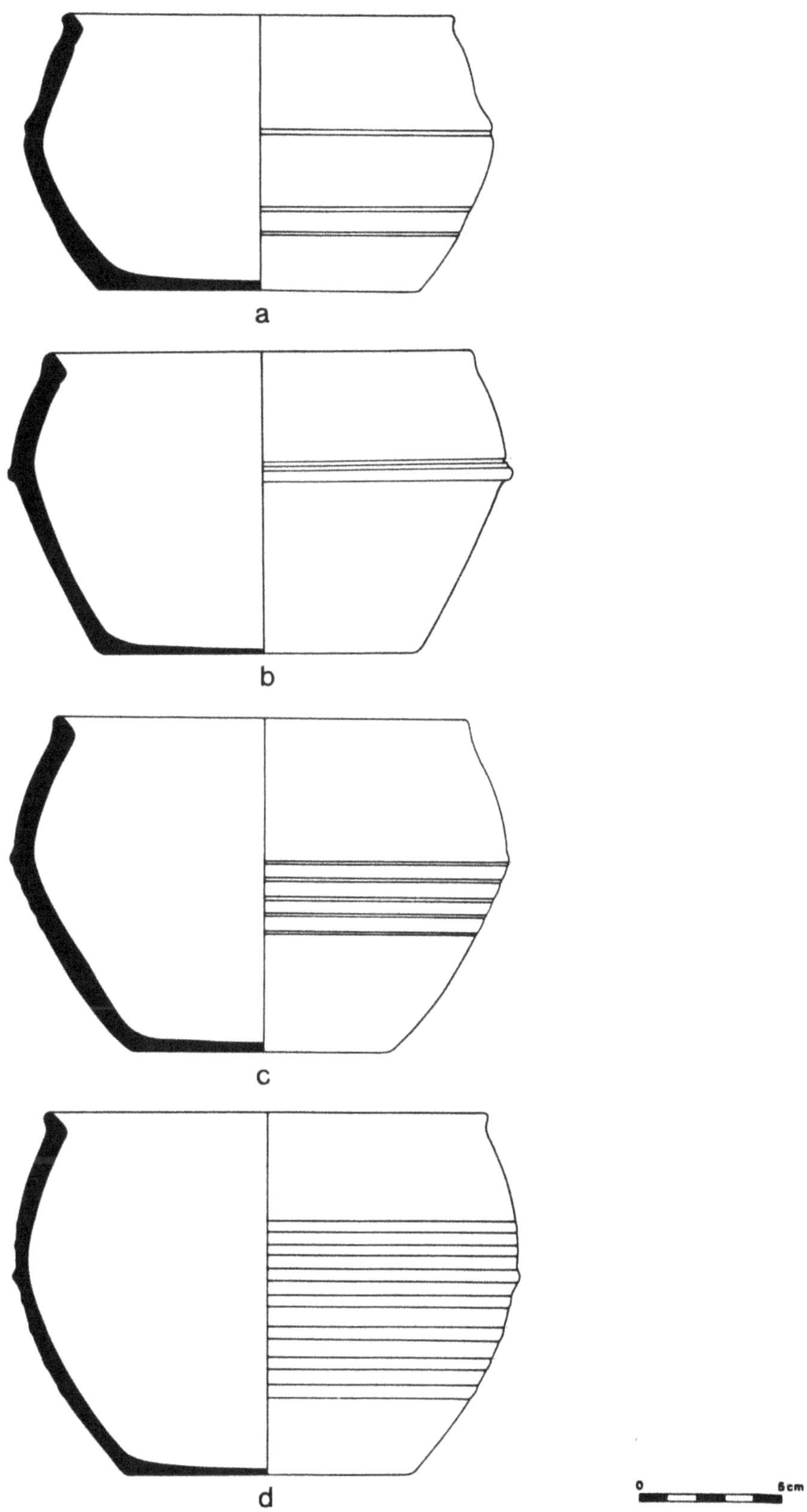


Figure 5.9. Carinated jars (Form 40).

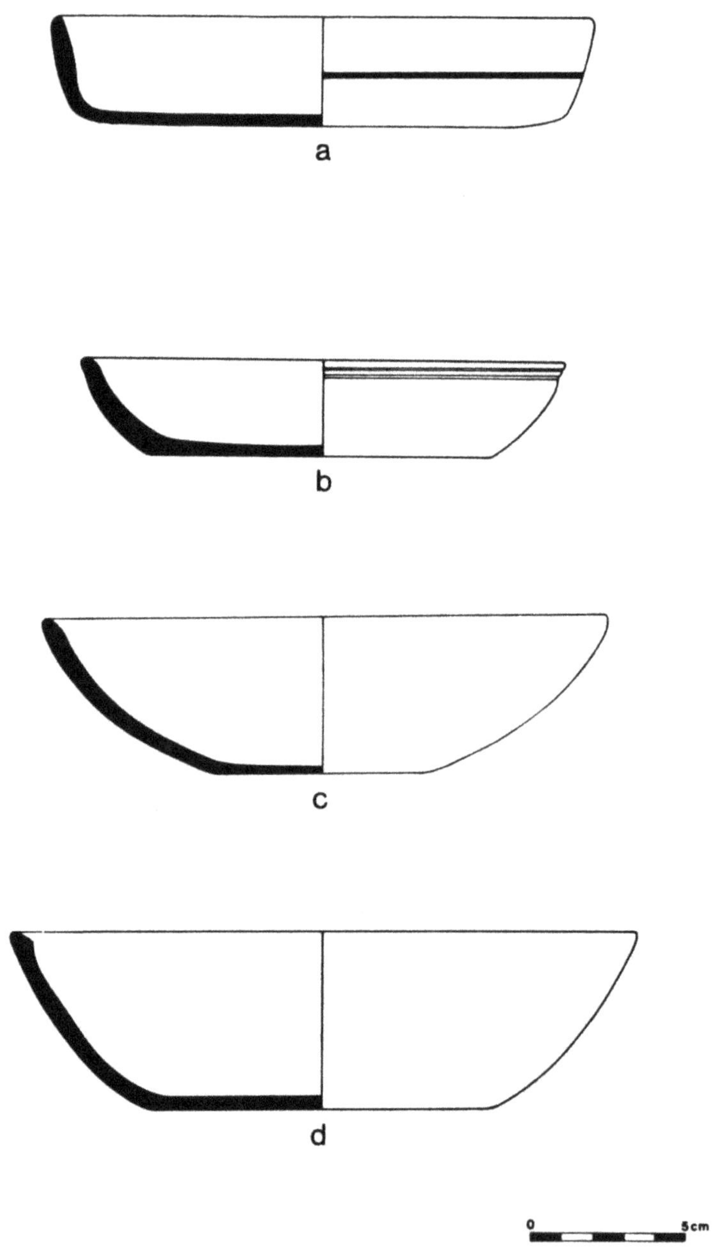


Figure 5.10. Shallow bowls. (a-c) Shallow bowl with simple rounded lip (Form 12). (d) Shallow bowl with bevelled lip (Form 13).

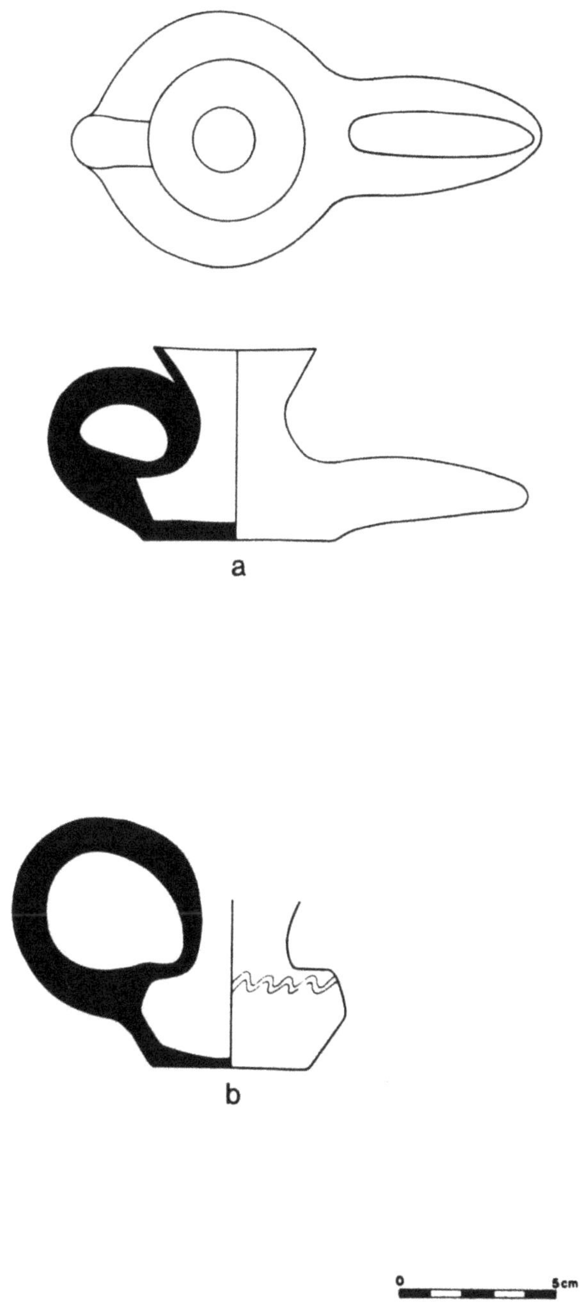


Figure 5.11. Oil lamps. (a) Lamp with protruding nose and handle attached below rim (Form 93). (b) Lamp with incised decoration under green glaze.

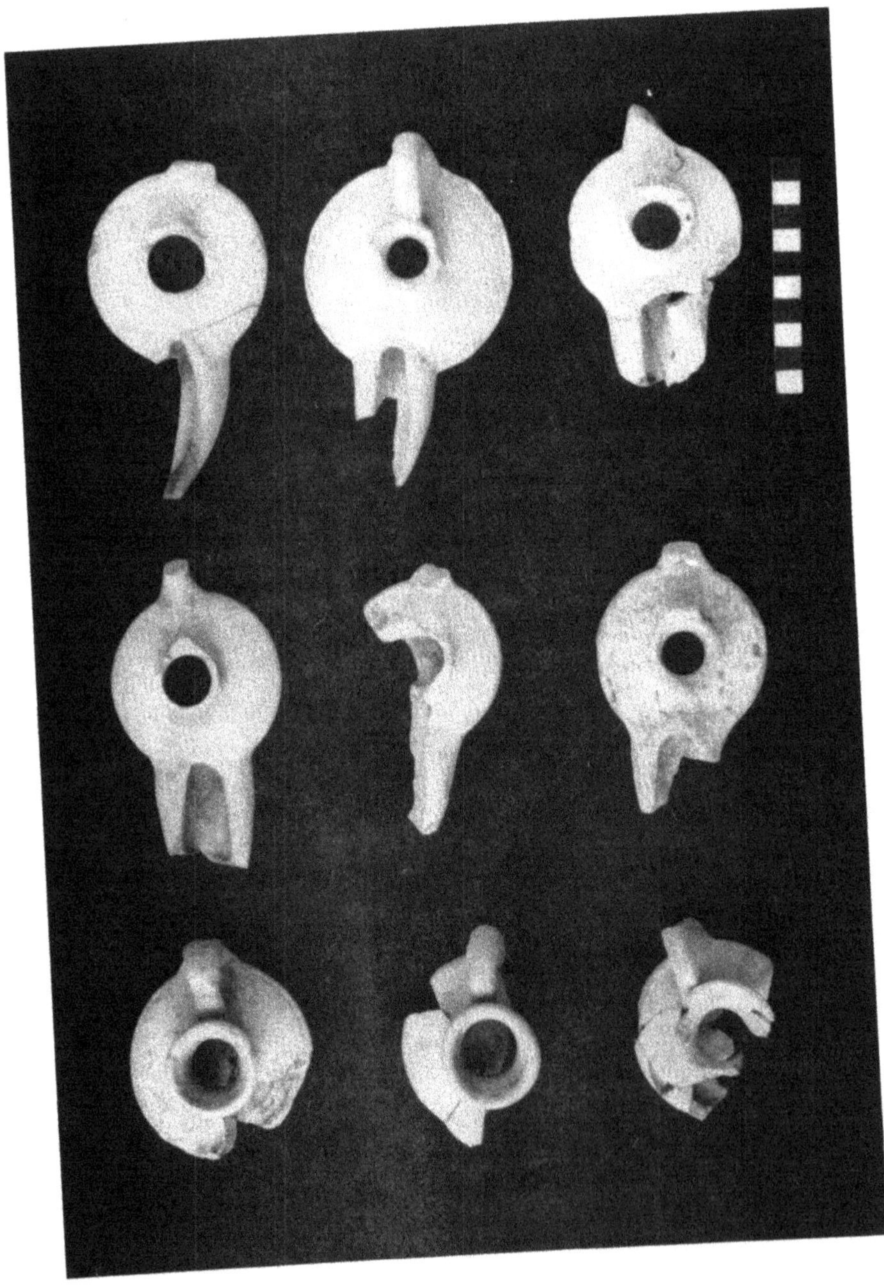

Figure 5.12. Nosed oil lamps.

## Cooking vessels

The most common cooking vessel (6.3%) is a shallow, handmade pot (Form 10). It is characterized by a simple lip, either rounded or slightly flattened, a finger- or tool-impressed strip applied just below the rim on the exterior, a slightly rounded body, and a flat to slightly convex base (Figure 5.1b). It is often charred on the exterior. A less common variety (1.8%) is a deep, handmade cooking pot (Form 11). It is characterized by a rounded or slightly flattened rim, vertical or inwardly sloping sides, usually with an impressed strip or horizontal lug below the rim, and a flat base (Figure 5.1c). Its exterior is also frequently charred.

## Minor forms

A number of other vessel forms found at al-Basra occur much more rarely at al-Basra. These include flasks with a variety of different rim shapes (Forms 61, 62, 68, 71, and 79: Figure 5.13a-b, g-h); small cups with vertical sides and flat bases (Form 110: Figure 5.13c-d); and a number of miscellaneous jars (Form 54: Figure 5.1a; Form 78: Figure 5.14c). There is also a small group of forms, some of them jars with bodies that slope inwardly, that are distinguished from the rest of the collection by their thick walls and brick red paste (Form 00: Figure 5.14a).

## Bases and handles

Unlike the rims, the variety of bases and handles is rather limited. Virtually all of the bases are flat, or slightly convex (Form 80). Although they range in size from about 5 to 15 cm in diameter, the most common size is about 10 cm. The bases typically have spiral grooves and ridges on the interior surface, made when the potter "opened" a ball of clay to form the vessel as the wheel rotated (Figure 5.15). They also have straight, parallel lines on the exterior surface (Figure 5.15), indicating that the vessel was removed with a string or cord from the wheel base while it was stationary (Figure 5.16) (Rye 1981:74-80). Ring, inset, flanged, and footed bases, which are usually found on glazed vessels, are present in very small quantities (Forms 82, 83, 84, 87).

Handles occur in two sizes. Small, relatively narrow handles seem to have been attached to pitchers and lamps. Large, thick handles appear to belong to large storage jars. A few small handles have a conical-shaped button on the top of the handle, probably to serve as a grip for the thumb; similar conical protuberances have been found on vessels at other Islamic sites in North Africa (Bel 1914: Figure 19; Golvin 1965:Plate 71; Zozoya 1980a:Figures 9, 14). Finally, there are a few examples of small handles made of two parallel coils attached to each other.

In addition to rims, bases, and handles, a number of other ceramic objects were found at al-Basra. One of the most prevalent was a small round disc (Form 100), generally about 3 cm in diameter, which appeared to be cut from broken pieces of pottery. These discs, which were found throughout the site, may have served as some sort of gaming object.

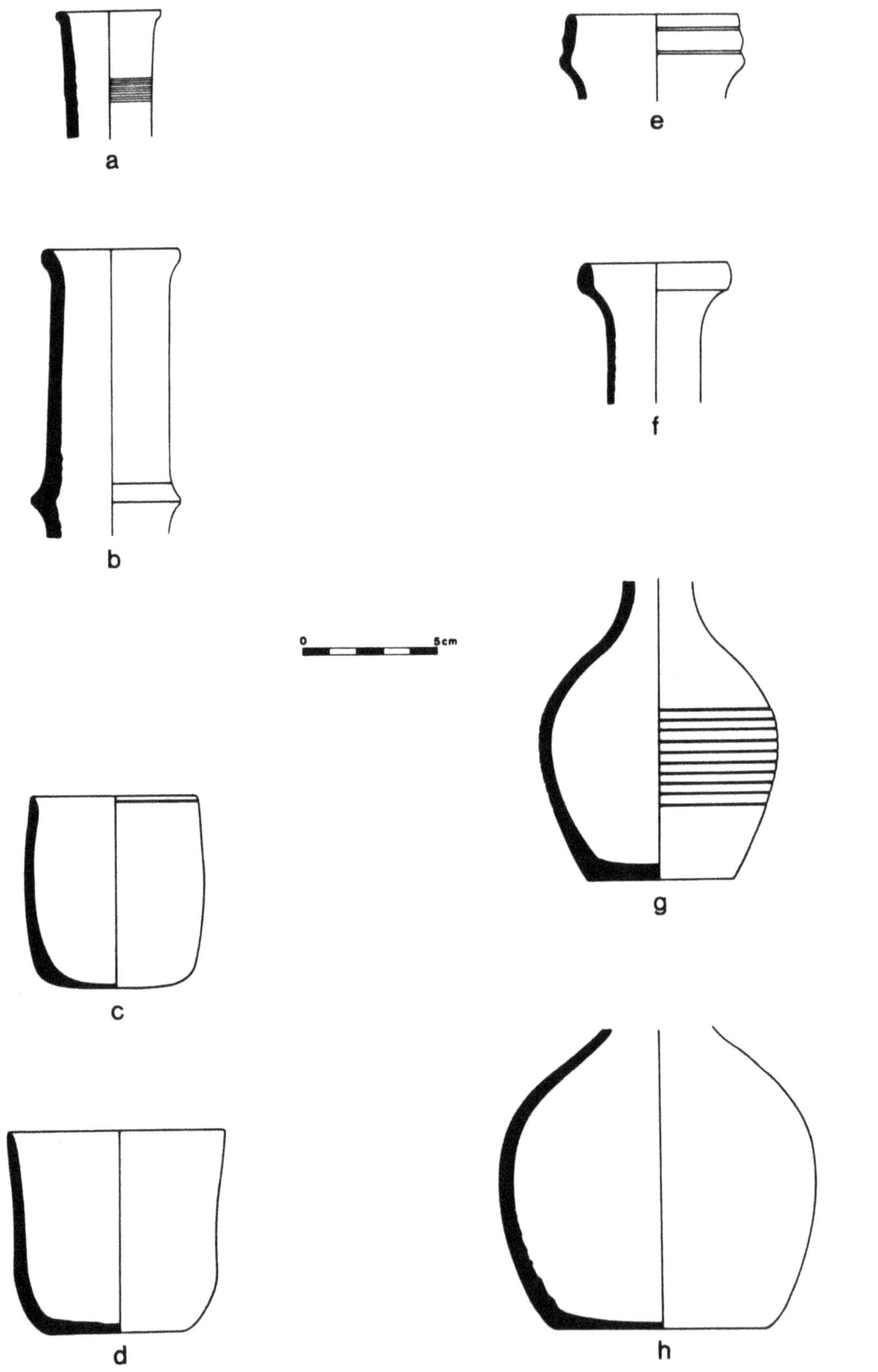


Figure 5.13. Flasks, cups, and squat jars. (a-b) Flasks with everted rims (Form 71). (c-d) Cups (Form 110). (e) Flask with grooved band rim (Form 79). (f) Flask with outset band rim (Form 68). (g) Squat jar with ribbed body (Form 61). (h) Squat jar (Form 62).

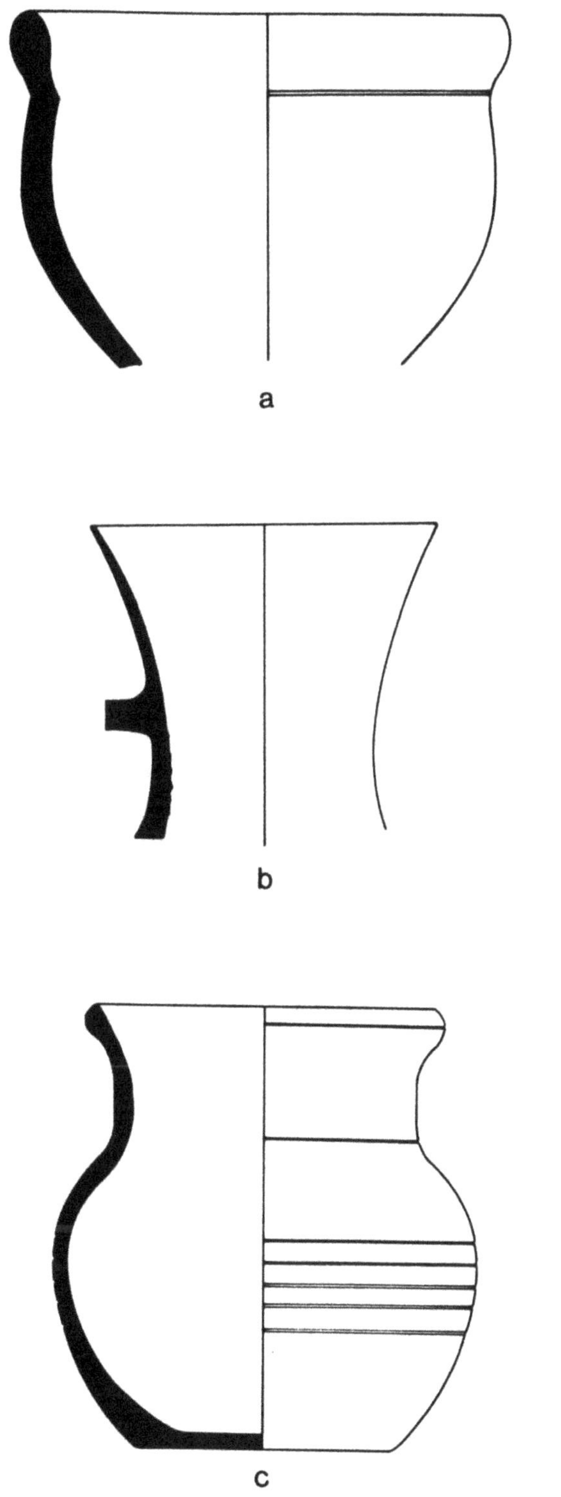


Figure 5.14. Miscellaneous jars. (a) Jar with band rim and thick, brick-red walls (Form 00). (b) Jar with tall, curved neck (Form 00). (c) Jar with triangular lip. curved neck, and ribbed body (Form 78).

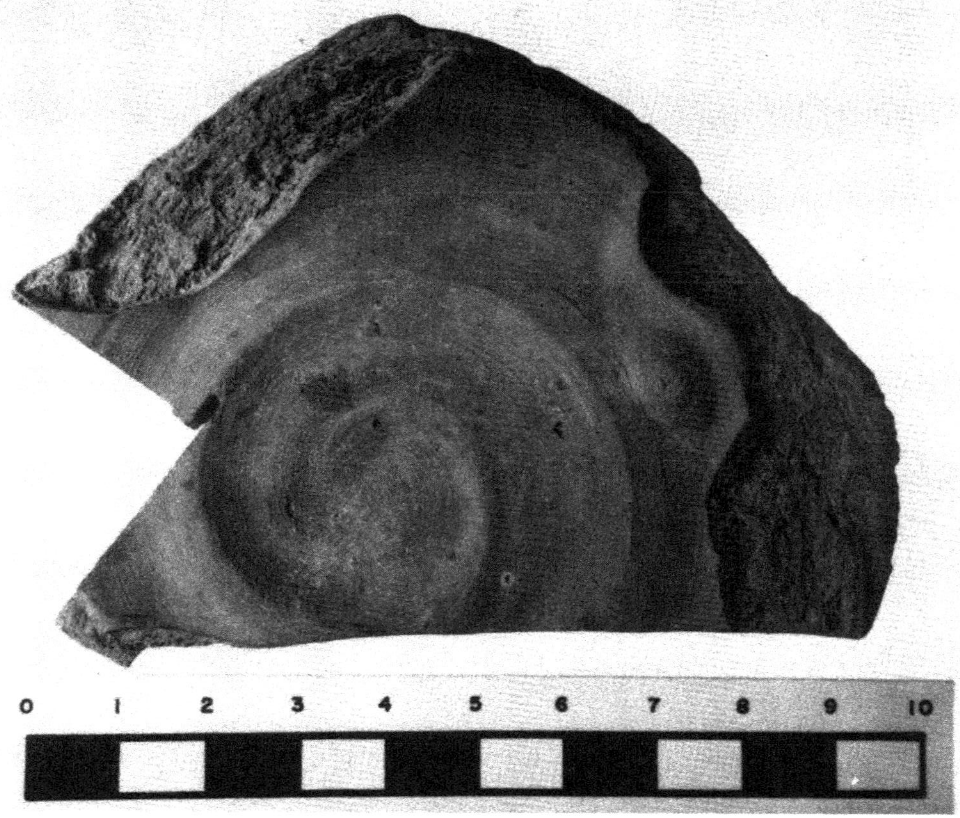


Figure 5.15. Interior of flat base, showing spiral grooves.

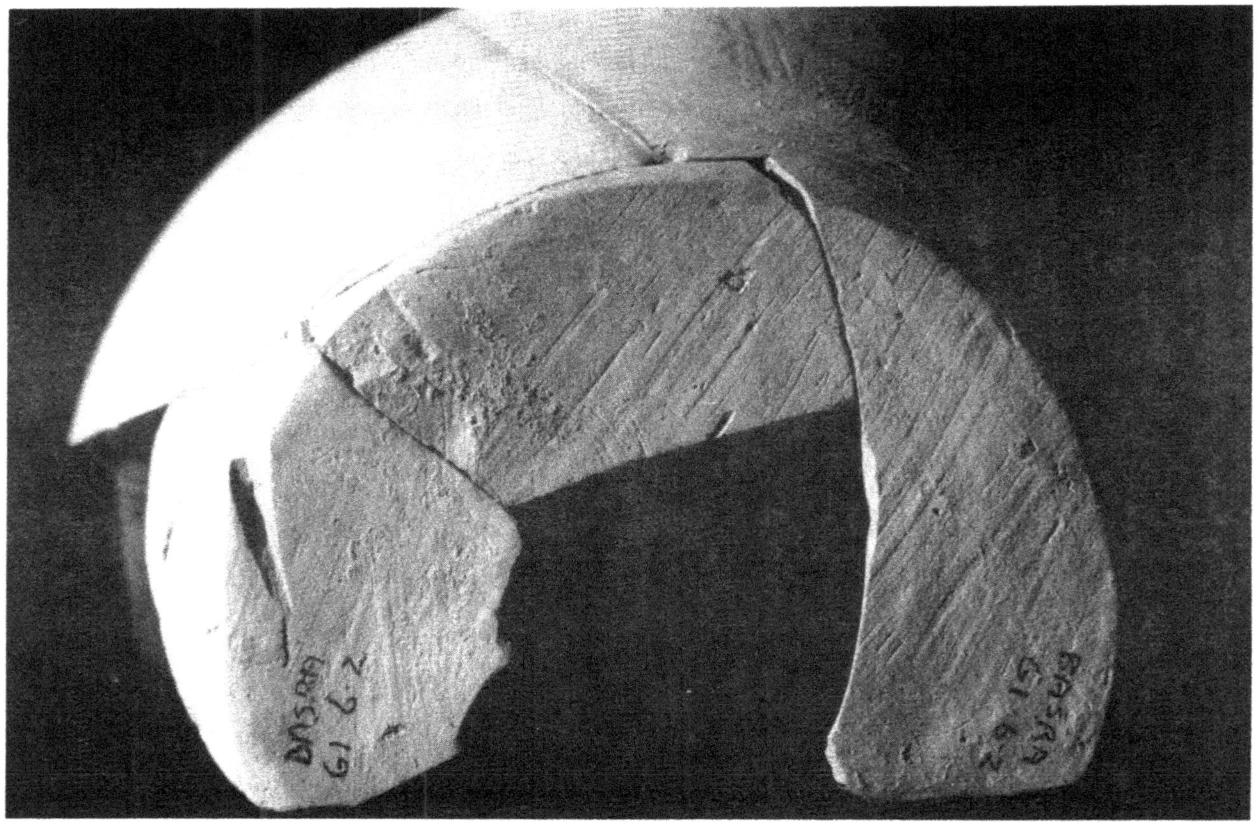

Figure 5.16. Exterior of flat base, showing drag lines made when vessel was removed from wheel-head.

Much less common were ceramic filters (Form 97) which had holes punctured through the clay with a sharp tool and which were placed in the neck of water storage jars in order to prevent dirt or insects from falling into the water; cylindrical spouts (Forms 90, 91); and small, round lids with button-like handles (Form 92).

## Decorative Techniques

The decorative techinques evident in the al-Basra collection consist of appliqué on handmade wares, and incising, *sigillé*, painting, and glazing on wheelmade wares.

At al-Basra, appliqué involved the application of a strip of clay to a handmade vessel; the clay strip was then pressed with a tool or finger to create a decorative pattern (Figure 5.1b, c).

The most common decorative technique present on unglazed pottery at al-Basra is incising (described as *décor gravé* or *décor trainé* by early French archeologists). A popular incised motif consists of a line or band of lines incised horizontally or in a wavy pattern around the neck or shoulder of a vessel (Figures 5.5b, 5.6, 5.7b, 5.17e). The lines are made with a pointed tool or a comb pressed into the soft clay as the vessel turned on a wheel; sometimes the bands are cross-hatched with diagonal lines. In addition to being decorative, incised lines may have served to conceal a seam in vessels that were made in several stages, particularly in the shoulders of large storage jars. Incised decoration is usually found on buff or coarse wares at al-Basra. Incising appears to be the most prevalent decorative technique found on unglazed utilitarian wares at Islamic sites in North Africa (Bel 1914; Golvin 1965).

A less common decorative technique found at al-Basra is *sigillé*, in which lines or motifs are cut into soft clay with a variety of pointed and sharp, flat tools (Figure 5.17a-c, d, f). The motifs generally consist of geometric patterns incised into the soft paste with a knife or other sharp instrument. The lines are parallel or cross-hatched. In some cases, the motif consists of a series of triangular shapes made by a tool pressed into the clay. *Sigillé* is a fairly common decorative technique present at other early Islamic sites in North Africa.

Another decorative technique occasionally found at al-Basra is painting. As we noted earlier, the paint is usually applied directly to the surface, although sometimes over a white slip. The paint generally consists of an iron-oxide base which turns various shades of red and brown during firing, depending on atmospheric conditions. The motifs are simple lines and dots, some of which may be epigraphic (Figure 5.2). Most painted decoration appears on cream wares, although it is occasionally found on buff wares.

Although not technically a decorative method, slipping with a white clay or self-slipping with a wet rag is a method that appears to have been used occasionally by al-Basra potters to finish the surfaces of their pottery.

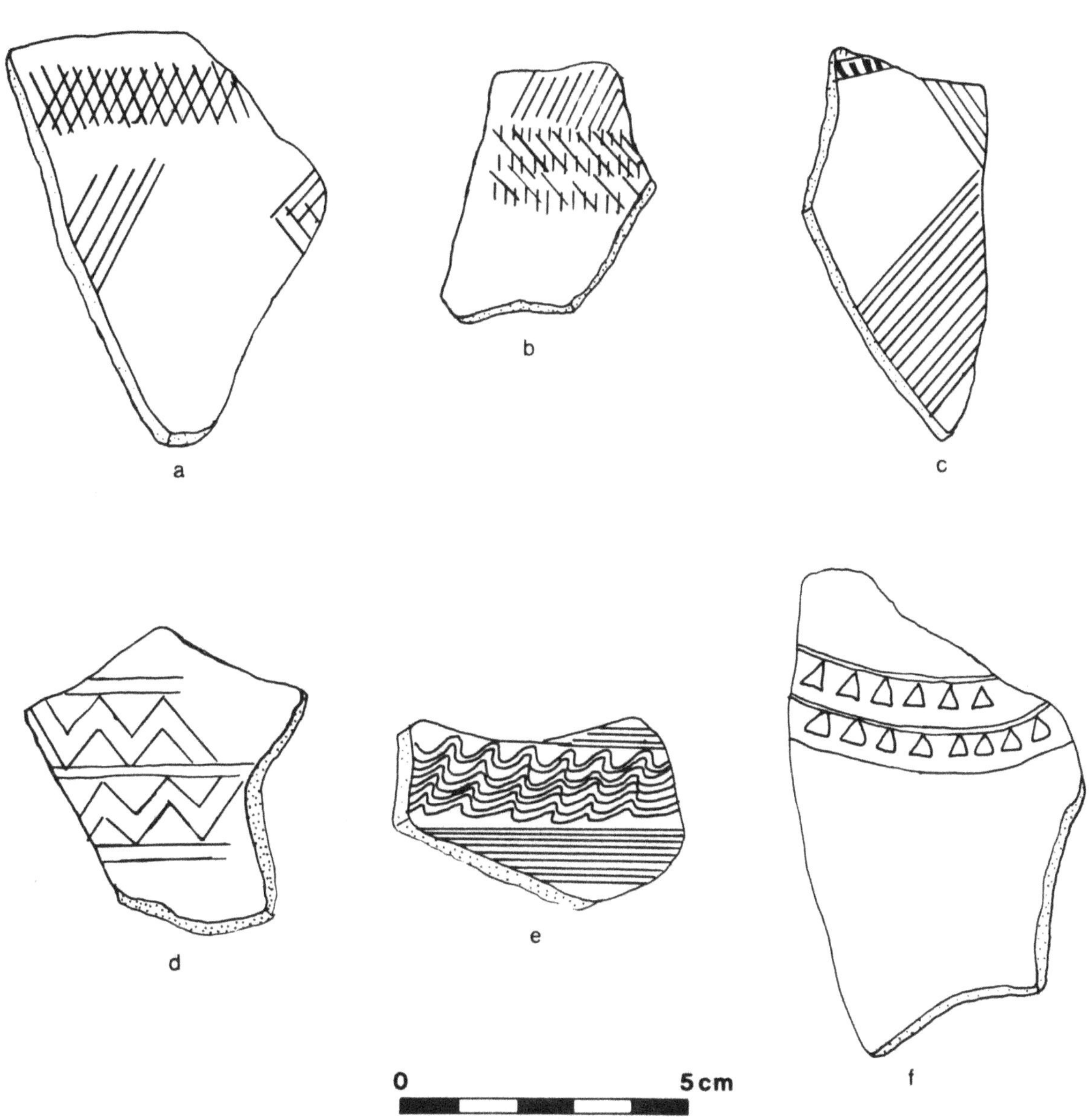

Figure 5.17. Incised and punctated pottery. (a-c) Sherds with lines incised or crosshatched into soft clay (<u>sigillé</u>). (d) Sherd with carved geometric motif. (e) Sherd with bands of incised horizontal and wavy lines (<u>décor gravé</u>). (f) Sherd with punctated decoration.

Glazing is another decorative technique evident in the al-Basra collection. With very few exceptions, however, this process does not seem to have been practiced at al-Basra. The composition of glazed wares appears to differ from the buff and cream wares that predominate at al-Basra. Only a few specimens, namely a green glaze painted base sherd (Figure 5.3f), seems likely to have been made at al-Basra.

While the decorative techniques found at al-Basra are relatively common throughout Islamic North Africa, a number of other popular techniques are completely absent from the al-Basra collection. As we will see in Chapter 7, the absence of pottery with stamped impressions and *sgraffito* (scratched) wares provides a way to tentatively date the al-Basra material.

## Faunal Remains

A large quantity of faunal material, amounting to nearly 600 kg by weight, was recovered during the al-Basra excavations. Animal bone, in fact, constitutes the second largest category of artifacts found at the site. Like the pottery, the material was collected and weighed by provenience unit and processed and stored in the laboratory in Tangier.

A small sample of bones was examined in 1981 in order to identity the types of animals that were present at al-Basra and to determine their relative proportion in the diets of city residents (Wattenmaker 1981). The study showed that domestic animals provided the major dietary contribution at al-Basra (Table 5.5). Goat and sheep appear to have been butchered most frequently, followed by cattle. Dogs and equids were present in the city, but do not seem to have been exploited for food. The following species were identified at al-Basra:

*Canis familiaris* (domestic dog) was represented by one bone which showed no signs of burning or butchering.

*Equus sp.* (equids) was represented by 14 bones, none of which exhibited evidence of butchering or burning, indicating that equids were probably not used as a food source. Equids are used in northern Morocco today for transport and as draft animals.

*Ovis aries* (sheep) was represented by 81 bones. Sheep are exploited in Morocco today for milk, wool, and meat.

*Capra hircus* (goat) was represented by 38 bones. Goats are raised for milk, meat, and hides in Morocco.

Some 797 bones were assigned to the general category of *Ovis-Capra* due to the similarity of sheep and goat bones. In addition, one bone was identified as *Ovis-Capra-Gazella*.

*Bos taurus* (cattle) was represented by 249 bones. Cattle are exploited for milk, meat, and hides and as draft animals in northern Morocco.

TABLE 5.5

COUNTS AND PERCENTAGES OF AL-BASRA ANIMAL BONE
(SELECTED SAMPLE)

| Taxon | Count (No.) | Percentage (%) |
|---|---|---|
| *Canis domesticus* | 1 | 0.0 |
| *Equus sp.* | 14 | 1.2 |
| *Sus scrofa* | 0 | 0.0 |
| *Capra hircus* | 38 | 3.2 |
| *Ovis aries* | 81 | 6.9 |
| *Ovis capra* | 797 | 67.5 |
| *Bos taurus* | 249 | 21.1 |
| Total | 1,180 | 100.0 |

## Skeletal Remains

The remains of two human skeletons were also found at the site, both of them inside the ancient walls. One appears to be a deliberate burial while the other seems to represent an accidental death or a secondary burial.

The partly articulated skeleton of one individual was found in the H1 sounding in the eastern part of the site. Located about 1 m below the surface, the skeleton lay flexed on its left side and faced eastward. Several flat rocks were placed directly above it. Although it is not clear when the body was buried, its position below the foundations of two nearby late-phase walls suggests that it may have occurred sometime prior to the final occupation period.

The disarticulated remains of a second individual were found between the preserved firing chamber and the outer wall of the F3 kiln in the western part of the city. The skeletal material was scattered in and under the mud-brick debris that resulted from the collapse of the kiln. Because of the position of the remains and the absence of charring on the bones, it is likely that the body was placed in the kiln after it ceased operating but before the floor collapsed. Redman (1983) has suggested that the individual may have died while repairing the kiln or that the bones of an already dead individual may have been thrown into what had become an inactive kiln.

Although it is possible that individuals were buried inside the city during its heyday, they were more likely interred in one of two cemeteries that existed outside the city during the early Islamic period and that are mentioned in the chronicles (al-Bakri 1965). One, which is still visible today, is situated about 100 meters west of the city, on the crest of a hill overlooking the Mda River valley. The other is located east of the city. The skeletons that were found inside the city walls thus may represent accidental deaths or burials that occurred during periods of destruction or abandonment at the site.

## Metal, Glass, and Stone Objects

A number of metal, glass, and stone artifacts were uncovered in the al-Basra excavations. Because most of the objects are fragmentary, their original form and function cannot be determined in many cases.

The bulk of the material consists of iron, either in the form of building nails or pieces of iron slag. In a few cases, the iron is shaped into rings. Copper artifacts are also relatively abundant. They generally consist of items of personal adornment, such as pins, finger rings, and ear rings, as well as a few thin copper leaves possibly made of hammered native metal. Lead objects are relatively rare, and include what appears to be a pair of riding stirrups.

Glass objects are fairly abundant. Although considerably fragmented, they consist primarily of bottle, vase, and/or lamp bases that are concave in shape where they were removed from the blowing instrument. Many of the surfaces are badly deteriorated from weathering and are characterized by an irridescent sheen. The fragments that retain their original color are green, blue, or yellow. Two small beads are also among the glass objects found at the site (Figure 5.18).

Stone objects occur very rarely. They include several fragments of large, round grinding stones and some probable flint tools.

Finally, three coins were found at the site (Figure 5.19). Two silver coins are believed to date to the early Islamic period but they have not been identified because of their badly worn condition. One was found on the surface of a plaster floor in the G3 sounding and the other was found in the middle levels of the F5 sounding. A third coin, made of copper, is clearly modern. It is characterized by a five-pointed Islamic star on one side and was discovered near the surface of the D1 sounding.

## Summary

The artifacts found at al-Basra encompass a wide variety of cultural material used in the construction of the city and in the everyday life of its inhabitants. Although the material is derived from a comparatively small portion of the site, it is assumed to be fairly representative of the total, not only because of its large quantity but because it comes from soundings distributed across the site.

The pottery, which constitutes the largest category of artifacts recovered at the site, encompasses a wide variety of wares and forms.

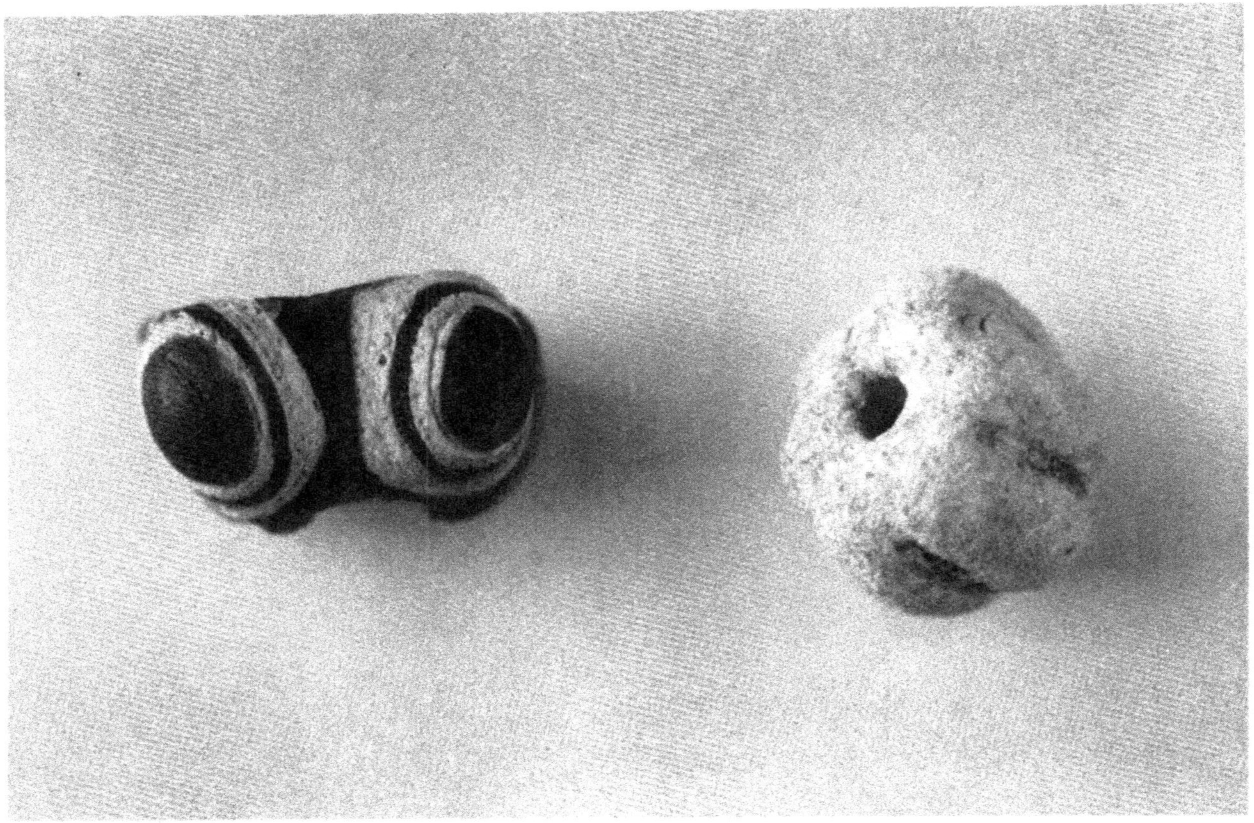

Figure 5.18. Beads from al-Basra.

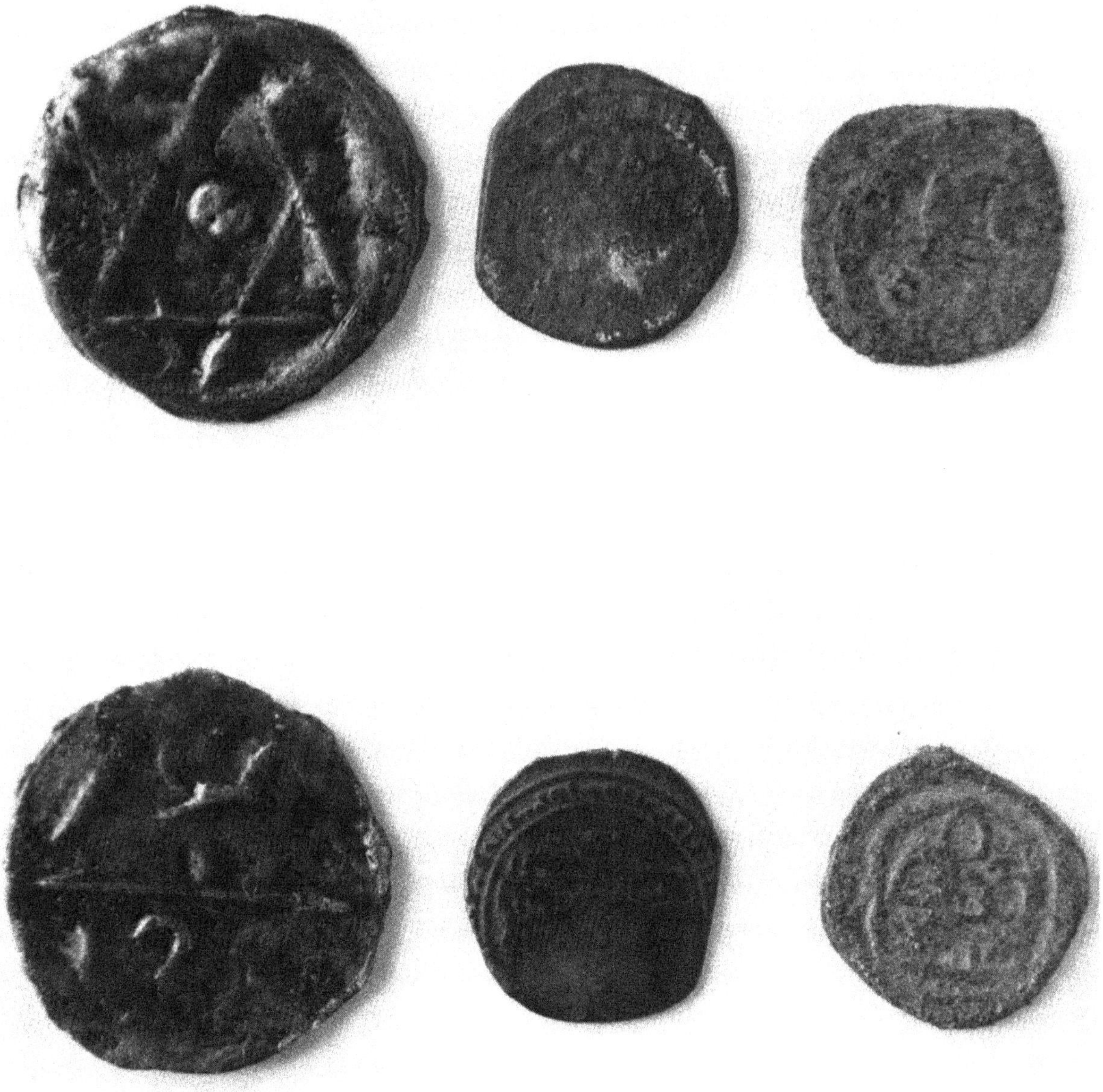

Figure 5.19. Coins from al-Basra. Top and bottom, from left: modern coin with five-pointed star; unidentified coin found in Phase 3 residential area context; unidentified coin found in Phase 3 industrial area context.

The bulk of the pottery collection, however, consists of only a few vessel types, which were used primarily for the preparation and storage of food and liquids and for artificial lighting. The collection contains very few examples of ceramic or other material that may have been used solely for decorative purposes, such as the elaborately decorated glazed plates or tiles found at other early Islamic sites in the western Mediterranean. It is possible that the excavations at al-Basra, because of their preliminary nature, failed to uncover areas of the city where elite residences were located and where luxury goods would be concentrated. On the other hand, it is possible that the excavations recovered the remains of some of these areas, particularly in the G soundings, but that they yielded relatively modest objects, reflecting the provincial nature of the town. Only additional work at the site will resolve this question. What is remarkable at this stage in our research is the <u>overall</u> homogeneity of the al-Basra pottery collection--with a limited variety of wares and forms dominating the total assemblage.

# CHAPTER 6

# DEPOSITIONAL ENVIRONMENT AT AL-BASRA

Before moving on to chronological issues, it is important to consider the processes by which pottery, animal bone, and other artifacts found their way into the archeological record at al-Basra. These processes include both cultural activities, such as trash disposal and building construction, and noncultural actions, such as wind and water erosion. As several researchers have shown (e.g., Schiffer 1976, 1983), depositional processes can have a considerable effect on the horizontal and vertical distribution of artifacts across a site.

In investigating some of these issues, this section (1) describes the types of depositional processes that are likely to be involved in the formation of an urban site, (2) examines the characteristics of artifacts, i.e., size, density, and type, present in different types of deposits at al-Basra, and (3) identifies residual or badly disturbed deposits that could affect the chronological ordering of artifacts at al-Basra.

## Types of Depositional Processes

The processes involved in moving artifacts from their original, or systemic, context to their final position in the archeological record are particularly complex at urban sites where material seldom remains where it is used, dropped, or discarded (Boone 1980; Schiffer 1976, 1983; Wright 1981). A broken water jar, for example, may be removed from a kitchen where it was used and then discarded in an abandoned lot on the periphery of the city; at a later time, the material in the lot may be reused as fill in the construction of a new building; the broken jar becomes part of the archeological record when the building collapses. Thus, the archeological context of the vessel varies considerably--both spatially and temporally--from its original place of use as a result of human intervention.

Several processes are generally at work in the formation of an urban archeological site (Schiffer 1976, 1983; Wright 1981). Primary deposition involves the loss or discard of artifacts at their place of use. As we mentioned above, this process rarely occurs at urban sites because broken or unusable items are usually removed from habitation areas and discarded elsewhere. However, it does occur when sites have been abandoned rapidly or destroyed cataclysmically, such as Pompeii or Akroteri (Thera). Artifacts found at their place of use serve as reliable indicators for spatial and temporal studies.

Secondary deposition involves the collection of broken and/or unusable items from their place of use and their subsequent discard as secondary trash elsewhere. This process is probably responsible for the deposition of most cultural material at urban sites. As studies of disposal practices at other urban sites show (Boone 1980; Brewster-Wray n.d.), broken or unuseable items are commonly discarded in abandoned buildings or empty lots by city residents who do not have access to modern trash disposal services. Cultural material that is discarded as secondary garbage serves as a fairly reliable chronological index but a relatively poor spatial marker.

Tertiary deposition involves the reuse or recyling of secondary trash for other purposes, such as construction fill or in building material. This type of depositional behavior is present at sites with substantial architecture and/or multiple building levels. The cultural material in tertiary deposits generally is not well suited for chronological or spatial studies because the material has been completely removed from its original systemic context.

Noncultural deposition involves the transportation and redeposition of artifacts by natural forces, such as wind and water erosion or animal activities. Although this type of depositional process is generally not a major factor at urban sites, it may be present to the extent that a site is subjected to flooding during its occupation or to wind and rain erosion after its abandonment. Cultural material in deposits formed by natural processes does not serve well as a chronological or spatial indicator because the material is almost entirely residual.

The characteristics of artifacts found in deposits formed by different processes--theoretically, at least--should vary. While a variety of attributes have been used in previous studies (see Schiffer 1983 for examples), I will concentrate on three attributes that seem to be most relevant to the al-Basra material. These are sherd size, sherd and bone densities, and the presence or absence of metal and glass objects.

The size of sherds found in archeological contexts may be directly related to the amount of disturbance and redeposition they undergo during their depositional lifetime (Bradley and Fulford 1975; Brewster-Wray n.d.). All other things being equal, the size of sherds decreases as the number of episodes of redeposition increases because the sherds continue to be subjected to breakage. For example, sherds found in secondary contexts should be smaller than those found *in situ*; those found in tertiary contexts should be smaller than those present in either primary or secondary situations. Sherd and bone densities in terms of weight can also be related to the depositional behavior of urban residents (Boone 1980; Brewster-Wray n.d.). The densities of both pottery and bone should be higher in dumps where they are deliberately discarded than on floors where they are periodically swept up and removed. The density of animal bone should be particularly low on occupation surfaces because of its noxious character. Finally, the presence or absence of metal and glass objects can also be associated with the depositional habits of city inhabitants which, in most cases, are determined by cultural attitudes toward the value and/or reuse potential of certain objects. At the medieval site of Ksar es-Seghir, for example, the occurrence of copper and glass fragments primarily in dumps rather than on floors suggests that these items may have had little reuse value or were highly noxious to local residents (Boone 1980).

## Depositional Processes at al-Basra

### Method of Study

The study of depositional factors at the site of al-Basra proceeded in the following way. Individual provenience units, i.e., level-loci, were assigned to specific depositional categories on the basis of stratigraphic evidence, including architectural features and sedimentary characteristics. This involved most of the 215 level-loci excavated at the site.

Average sherd size, sherd and bone densities, and the presence or absence of metal and glass were determined for each provenience unit for which such information was available. Average sherd size was calculated by dividing the total weight of sherds (in kilograms) by the total number of sherds in a provenience unit. Sherd densities were calculated by dividing the total weight of sherds (in kilograms) recovered from a provenience unit by the estimated total volume of earth (in cubic meters) removed from the unit. Bone densities were computed in the same way. The presence or absence of metal and glass objects in each provenience unit was determined from field records. Calculations were made for a total of 137 provenience units, including nearly all level-loci in deeply stratified soundings and a sample of level-loci in shallow soundings. This information is presented in Appendices 1 and 2.

Finally, the mean sherd size, the mean sherd and bone densities, and the number of provenience units containing metal and glass objects were determined for each depositional category. This information is summarized in Table 6.1 and Figure 6.1.

TABLE 6.1

AVERAGE SHERD SIZE, AND POTTERY AND BONE DENSITIES, BY DEPOSITIONAL CATEGORY

| Deposit Category | Total Units (No.) | Average Sherd Size (g) | Sherd Density ($k/m^3$) | Bone Density ($k/m^3$) |
|---|---|---|---|---|
| Sterile soil | 11 | 13.2 | 2.3 | 0.5 |
| Topsoil | 16 | 9.6 | 2.4 | 0.8 |
| Mud-brick collapse | 15 | 11.8 | 4.2 | 1.8 |
| Primary floor | 13 | 12.8 | 6.0 | 2.1 |
| Secondary floor | 5 | 17.2 | 5.1 | 2.4 |
| Kiln collapse | 13 | 15.7 | 4.6 | 1.8 |
| Gravel fill | 9 | 17.1 | 5.3 | 3.5 |
| Undifferentiated fill | 41 | 15.3 | 8.1 | 3.7 |
| Construction fill | 10 | 15.5 | 13.8 | 4.3 |

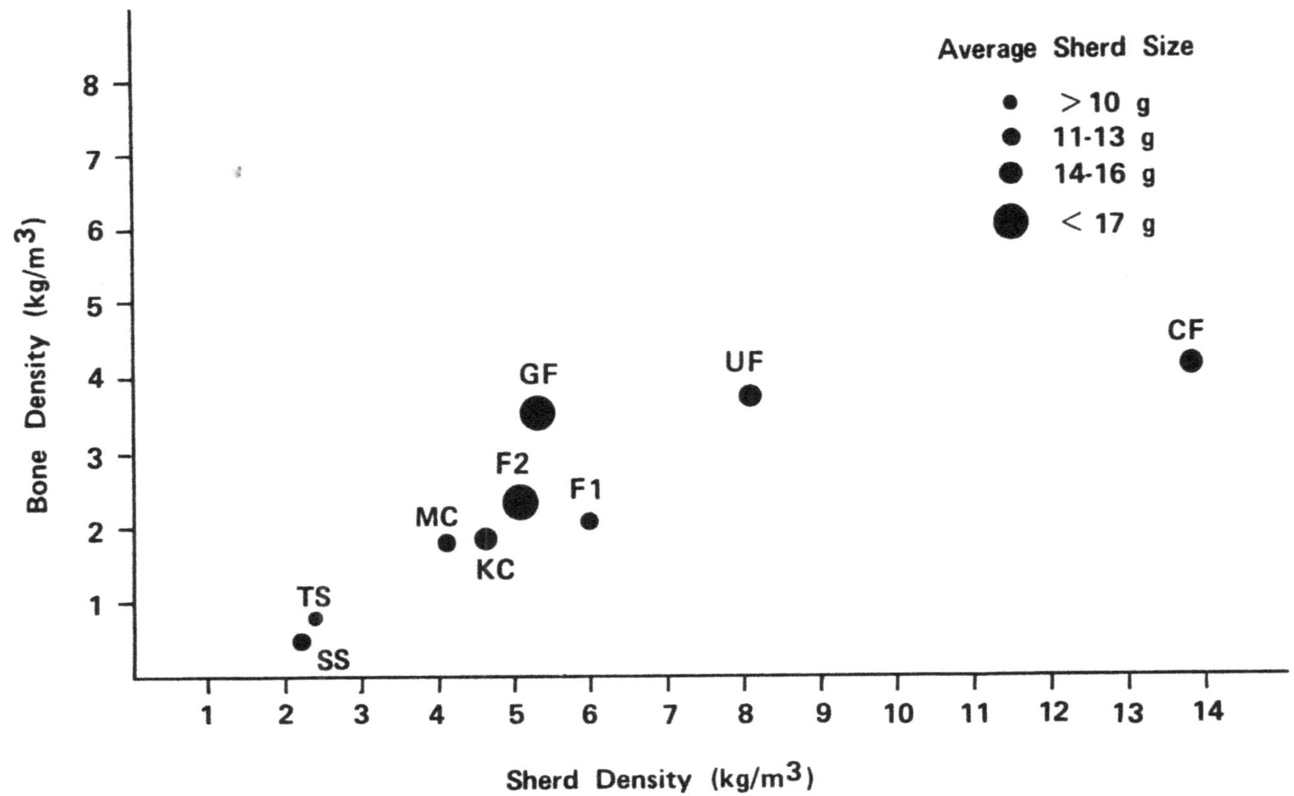


Figure 6.1. Distribution of depositional categories by average sherd size, sherd density, and bone density. Key: SS, sterile soil; TS, topsoil; MC, mud-brick wall collapse; KC, kiln collapse; F1, primary floor deposit; F2, secondary floor deposit; GF, gravel fill; UF, undifferentiated fill; CF, construction fill.

## Types of Deposits

The nine depositional categories prevalent at al-Basra are described below.

### Primary-floor deposits (F1)

Primary floor deposits occur directly above plaster, stone, or hard-packed earth surfaces. They are overlain with the debris of collapsed buildings, such as building stone, deteriorated mud brick, and broken roof tiles. In some cases, primary floor deposits contain the remains of pottery vessels that were apparently broken *in situ* by the weight of falling roofs and walls, e.g., G2 5-2/6-2, G2 5-1/6-1, and F3 3-0, or were left behind when the buildings were abandoned.

### Secondary-floor deposits (F2)

Secondary floor deposits occur directly above plaster, stone, or hard-packed earth surfaces. In contrast to primary deposits, however, they are not sealed from above by building collapse debris. Instead, they consist of a relatively thick layer of undifferentiated fill which appears to have been discarded over a relatively long period of time.

### Kiln-collapse deposits (KC)

These primary and secondary deposits occur in direct association with two updraft kilns found at the site. They contain the debris of collapsed kilns, including brick and roof tile fragments, and other material deposited as secondary fill in the area.

### Undifferentiated-fill deposits (UF)

These secondary deposits are found in association with stone walls and floors and in open areas without any associated architectural material. The deposits contain little or no building debris. They are characterized by sediments that are usually loose and sandy and, less frequently, clayey. The deposits appear to have accumulated as a result of numerous secondary dumping or discard episodes.

### Construction-fill deposits (CF)

These tertiary deposits are found adjacent to or directly below the base of stone walls or floors. They consist of fill--often including building stone fragments--that presumably accumulated in another part of the site and was subsequently reused for the foundation of a new building.

### Mudbrick-wall-collapse deposits (MC)

These tertiary deposits are usually found toward the upper part of stone walls. They represent the deteriorated mud brick used in the construction of the upper part of walls and the debris of fallen roofs. The deposits are distinguished by their orangish-red clayey matrix and numerous rooftile and charcoal inclusions. Other cultural material found in these deposits may have served as a tempering material in the manufacture of unfired mud brick.

Gravel-fill deposits (GF)

These tertiary deposits contain large quantities of gravel that appear to have been deposited by fluvial action. They are found in association with stone walls and in areas with no architecture. The material in the deposits is very likely residual and was redeposited along with the gravel by water action.

Topsoil deposits (TS)

These tertiary deposits are found at the surface of the site and range in depth from a few centimeters on the hill slopes to half a meter on the valley floor. They accumulated as a result of soil erosion from the hills around the site. The cultural material in these deposits has moved a considerable distance over a period of several centuries.

Sterile-soil deposits (SS)

These deposits are located below the cultural levels at the site. They consist of dark gray to black calcareous clayey sediments in the valley and pebbly red clayey sediments on the hillsides. They contain a small amount of cultural material that has filtered down from the habitation levels above.

## Characteristics of Deposits

### Average sherd sizes

As Figure 6.1 illustrates, artifact sizes and densities vary from one depositional category to another. Although they follow the patterns we expected in most cases, they diverge in others.

The average sherd size is relatively small in topsoil, sterile-soil, and mud-brick-wall-collapse deposits, all of which represent tertiary environments. The sherds in these deposits are likely to have been reduced in size by plowing, trampling, alluvial activities, or recyling as building material. The sherd size is unexpectedly small in some primary-floor deposits. Although some of these deposits contained vessels that appeared to have been broken *in situ* when the roof collapsed and thus were characterized by large, often refittable sherds, such as those found in G2 5-2/6-2 (see Figure 4.4), other primary floor deposits seem to have contained material that was discarded as secondary trash or embedded in the floor as a result of trampling. The variation in average sherd size in primary floor contexts suggests that a number of different factors were operating and, consequently, such deposits are not easily categorized.

The average sherd size is relatively large in undifferentiated fill, secondary-floor, and kiln-collapse deposits, all of which represent secondary deposits formed when material was moved from its place of use and redeposited in a dump or trash area. Surprisingly, large average sherd sizes are also found in gravel and building fill, which are tertiary deposits and would be expected to have fairly small sherd sizes because of frequent disturbance. The fact that they do not raises some questions. Sherds found in gravel deposits, which were almost always located in the lowest part of the valley floor, may have actually been subjected to relatively brief episodes of alluvial disturbance, especially in comparison with the debris in topsoil deposits that experienced several centuries of redeposition. The brief period of disturbance may not have been long enough or constant enough to reduce the sherds in size. Pottery found in construction fill may have been selected for its size, since larger fragments would fill an area more quickly. Sherds from waster dumps, for example, may have been used for building fill, although the material found in construction fill does not exhibit an unusually high proportion of bloated or overfired pieces.

In summary, the variation in average sherd sizes in different kinds of depositional environments at al-Basra seems to be partly a function of the amount of disturbance and redeposition that the material underwent during its depositional history. Average sherd size may also be affected by several other factors, such as the _nature_ of the disturbance, e.g., trampling versus roof collapse, and the _duration_ of the disturbance, e.g., ten years versus several centuries. It may also be a function of the _character_ of the pottery itself. For example, the relatively hard, kiln-fired pottery found at al-Basra may be more resistant to breakage and wear than most low-fired ceramics, making it a less sensitive indicator of depositional processes.

## _Sherd and bone densities_

Sherd and bone densities also vary among depositional categories (Figure 6.1). As was expected, artifact densities are relatively low in topsoil and sterile-soil deposits; neither type of deposit is associated with actual habitation levels at the site. On the other hand, sherd and bone densities are moderately low in primary- and secondary-floor, mud-brick-collapse, and kiln-collapse deposits, although unusually high bone densities occur in gravel deposits. The deposits in most of these cases are associated with working or living areas, in which the amount of debris would be expected to be kept to a minimum. Artifact densities are moderately high in undifferentiated fill deposits where cultural material was deliberately discarded. Finally, sherd and bone densities are highest in construction-fill deposits, in which cultural debris may have been purposely added in order to build up areas of the site for new construction.

Researchers at other urban sites have discovered that artifact densities generally seem to be lower in floor deposits than in secondary-fill deposits, which accumulate as a result of deliberate trash disposal (Boone 1980; Brewster-Wray n.d.). At al-Basra, however, sherd and bone densities in secondary-floor deposits, which were assumed to be loci of secondary trash disposal, were unusually low in comparison with general fill deposits. This suggests that trash accumulated on abandoned floors at a slower pace than it did at other secondary trash locations, perhaps because floors, especially those located inside buildings, were relatively inaccessible to urban residents for dumping.

## Presence of Metal and Glass Objects

As Table 6.2 shows, the presence of metal and glass objects also varies in relation to depositional type. Iron, as a category that includes nails and slag, appears frequently in most types of deposits, particularly in undifferentiated fill, gravel fill, kiln collapse, and topsoil deposits. Iron nails as a separate category, however, occur far less frequently in these deposits. Instead, they are found primarily in floor deposits and in mud-brick wall debris, both of which are associated with architectural material. This suggests that they were used in wall or roof construction and dropped into floor deposits when the structure collapsed. Interestingly, a similar concentration of nails in floor deposits was found in the Portuguese levels at Ksar es-Seghir (Boone 1980). Like nails, copper objects also were concentrated in floor, mud-brick, and topsoil deposits at al-Basra, suggesting that they too may have been used in building construction. Those copper objects that consisted of items of personal adornment may have been abandoned in rooms before the buildings collapsed. Glass fragments occur at al-Basra more frequently in construction-fill, secondary-fill, kiln-collapse, and floor contexts. The distribution of copper and glass objects differs somewhat from that at Ksar es-Seghir where both kinds of items were found mainly in dumps. The differences may reflect, among others, cultural differences concerning the value of copper and glass among the Portuguese or Islamic populations of these cities.

## Residual and Disturbed Deposits

As a result of the study, several types of deposits at al-Basra were identified as potentially residual and badly disturbed. These were deposits that consisted of topsoil, building fill, and the debris of collapsed mud brick. In the course of developing a pottery chronology for al-Basra (see Chapter 7), it became evident that a few of these deposits contained pottery material that was clearly mixed, i.e., the proportions of pottery wares and forms differed significantly from those in earlier or later deposits. On the basis of both stratigraphic and depositional evidence, the pottery data for these provenience units were not used in subsequent analyses that required temporally coherent assemblages.

The clearly residual pottery in topsoil deposits was treated as post-urban material (Phase 4) for later chronological study. The ceramic material from certain building-fill deposits that appeared to be badly mixed, e.g., G1 10-1, was not used in further analyses; however, the pottery from building-fill deposits that fell into the normal pottery sequence was retained. The pottery in mud-brick-wall deposits, which was also expected to be badly mixed, in fact closely reflected the ceramic material found in associated floor deposits, suggesting that some of it may originally have been placed on floors or that the mud bricks were tempered with contemporaneous ceramic material. As a result, the pottery in these deposits was considered in the chronological study.

In addition to identifying disturbed deposits at al-Basra, the study underscores the complex and often particularistic nature of depo-

TABLE 6.2

COUNTS AND PERCENTAGES OF PROVENIENCE UNITS
WITH METAL AND GLASS OBJECTS,
BY DEPOSITIONAL CATEGORY

| Deposit Category | Total Units | Slag | | Iron | | Nails | | Copper | | Glass | | Lead | |
|---|---|---|---|---|---|---|---|---|---|---|---|---|---|
| | (No.) | (No.) | (%) | (No.) | (%) | (No.) | (%) | (No.) | (%) | (No.) | (%) | (No.) | (%) |
| Sterile soil | 6 | 1 | 17.0 | 0 | 0.0 | 1 | 17.0 | 1 | 17.0 | 3 | 50.0 | 0 | 0.0 |
| Topsoil | 9 | 1 | 11.0 | 5 | 56.0 | 0 | 0.0 | 3 | 33.0 | 0 | 0.0 | 0 | 0.0 |
| Mudbrick walls | 23 | 1 | 4.0 | 8 | 35.0 | 2 | 9.0 | 8 | 35.0 | 4 | 17.0 | 0 | 0.0 |
| Primary floor | 21 | 0 | 0.0 | 6 | 29.0 | 4 | 19.0 | 5 | 24.0 | 5 | 24.0 | 1 | 5.0 |
| Secondary floor | 5 | 0 | 0.0 | 2 | 40.0 | 1 | 20.0 | 1 | 20.0 | 1 | 20.0 | 0 | 0.0 |
| Kiln collapse | 16 | 1 | 6.0 | 8 | 50.0 | 0 | 0.0 | 3 | 19.0 | 4 | 25.0 | 0 | 0.0 |
| Gravel fill | 8 | 1 | 13.0 | 5 | 63.0 | 0 | 0.0 | 1 | 13.0 | 0 | 0.0 | 1 | 13.0 |
| Undifferentiated fill | 46 | 4 | 9.0 | 20 | 43.0 | 4 | 9.0 | 8 | 17.0 | 10 | 22.0 | 0 | 0.0 |
| Construction fill | 11 | 0 | 0.0 | 3 | 27.0 | 0 | 0.0 | 1 | 9.0 | 7 | 64.0 | 0 | 0.0 |

sitional processes at urban sites. It suggests that a number of criteria, such as average sherd size and artifact densities, may be usefully examined in depositional studies. It also emphasizes the importance of supplementary evidence in understanding the specific nature of depositional processes involved in creating the archeological record at an urban site.

# CHAPTER 7

# AL-BASRA POTTERY CHRONOLOGY

The development of a ceramic chronology for the early Islamic site of al-Basra is important for a number of reasons. First of all, it provides a temporal framework for a study of the pottery industry at al-Basra (see Chapter 8). With relatively secure chronological controls, it is possible to monitor changes in the organization and technology of pottery production at the center during its 300-year history. The development of a ceramic sequence also sets the stage for future excavations at al-Basra and for archeological surveys in northern Morocco. With a pottery sequence established, it will be possible to date other architectural and artifactual material excavated at al-Basra and identify other Islamic sites in the surrounding region. Finally, the development of a ceramic chronology provides an important contribution to the comparative study of early Islamic pottery in the western Mediterranean. The al-Basra pottery sequence represents one of the few chronologies available for the early period of Arab settlement in North Africa and Spain that is based on firm stratigraphic evidence and systematic sampling procedures.

The chronological evidence is presented in this chapter in three sections: (1) a discussion of the methodology used to develop the sequence, (2) a description of the chronology itself, and (3) a comparative study of ceramic material from other early Islamic sites in Morocco, Algeria, Tunisia, and Spain.

## Methodology

Because of the availability of abundant stratigraphic evidence at al-Basra, the method I used to develop a pottery chronology was fairly simple and straightforward. It involved calculating the relative frequencies of wares and forms for each provenience unit and ordering the units into a sequence according to their stratigraphic position within individual soundings.

The wares and forms selected for the analysis consisted of those that appeared to be particularly sensitive to temporal change. These were all of the ware categories (sandy buff, fine buff, and very fine buff, cream-colored, coarse, glazed, and handmade) and six major form categories (40, 41, 42, 43, 48, and 93). Although a few minor forms (those amounting to less than 2% of the collection) also exhibited some temporal variation, they were not included in the major form analysis because of their small quantities. They are, however, briefly discussed in a separate section of this chapter. In all, more than 5,000 rim sherds, identified by ware and form, were included in the chronological study.

The provenience units used in the analysis included nearly all of those from deeply stratified soundings (F1, F3, F5, G1, and G2) and a selection of others from minor soundings (A1, C1, D1, E1, F2, F4, F5, G3, H1, and I2). Although every effort was made to keep the smallest possible units of analysis intact in order to monitor gradual temporal changes, a few provenience units with low sherd counts were combined with adjacent units. A total of 133 provenience units were considered in the study.

The relative frequencies of wares and forms for each provenience unit in individual soundings are displayed in the bar diagrams in Figures 7.1-7.14. The actual counts are presented in Appendices 3 and 4. In the bar diagrams, the columns represent wares and forms and the rows represent specific provenience units (level-loci). Provenience units are displayed very much as they appeared stratigraphically, with units grouped by locus and arranged by level from top to bottom. Levels that represent the early occupation phase (Phase 1) are at the bottom; those that represent the post-urban phase (Phase 4) are at the top.

A visual inspection of the bar diagrams indicates that the relative frequencies of some wares and forms change in a consistent manner from early to late assemblages. The diagrams also show that the relative abundances of some wares and forms change in a similar way from one sounding to another, thereby allowing provenience units from different parts of the site to be correlated on the basis of ceramic evidence. Finally, the diagrams demonstrate that major shifts in the presence or relative abundances of certain wares and forms coincide with important architectural changes, thus allowing pottery assemblages to be assigned to specific occupation phases.

Minor forms and glazed and painted pottery are not included in the bar diagrams but they are discussed in a later section. Some of these relatively rare pottery types serve as potential diagnostic markers for certain occupation phases at al-Basra.

## Temporal Distribution of Wares

As Figures 7.1-7.7 show, the relative abundances of several al-Basra wares increase or decrease in a consistent way through time. The most striking changes occur in four of the seven ware groups: sandy buff, very fine buff, cream-colored, and glazed. Most of these changes seem to be the result of technological innovations in pottery production at the urban center. In the case of imported glazed wares, however, the changes appear to be the result of shifts in trading practices between the merchants of al-Basra and other urban centers. In the following discussion, the early occupation phase refers to Phase 1, the middle occupation phase to Phase 2, the late occupation phase to Phase 3, and the post-urban phase to Phase 4.

### Sandy wares

Sandy buff-firing wares are concentrated in the lowest levels at al-Basra, where they are associated with architectural remains from the early occupation phase. The coarsely textured wares are relatively abundant in the lowest levels of the residential soundings (G1 and G2 [Figures 7.4-7.5]) and in one of the kiln soundings (F3 [Figure 7.2]), where they range from 20% to 60% of each assemblage. Sandy wares decline in relative frequency in the second and third occupation phases at the site.

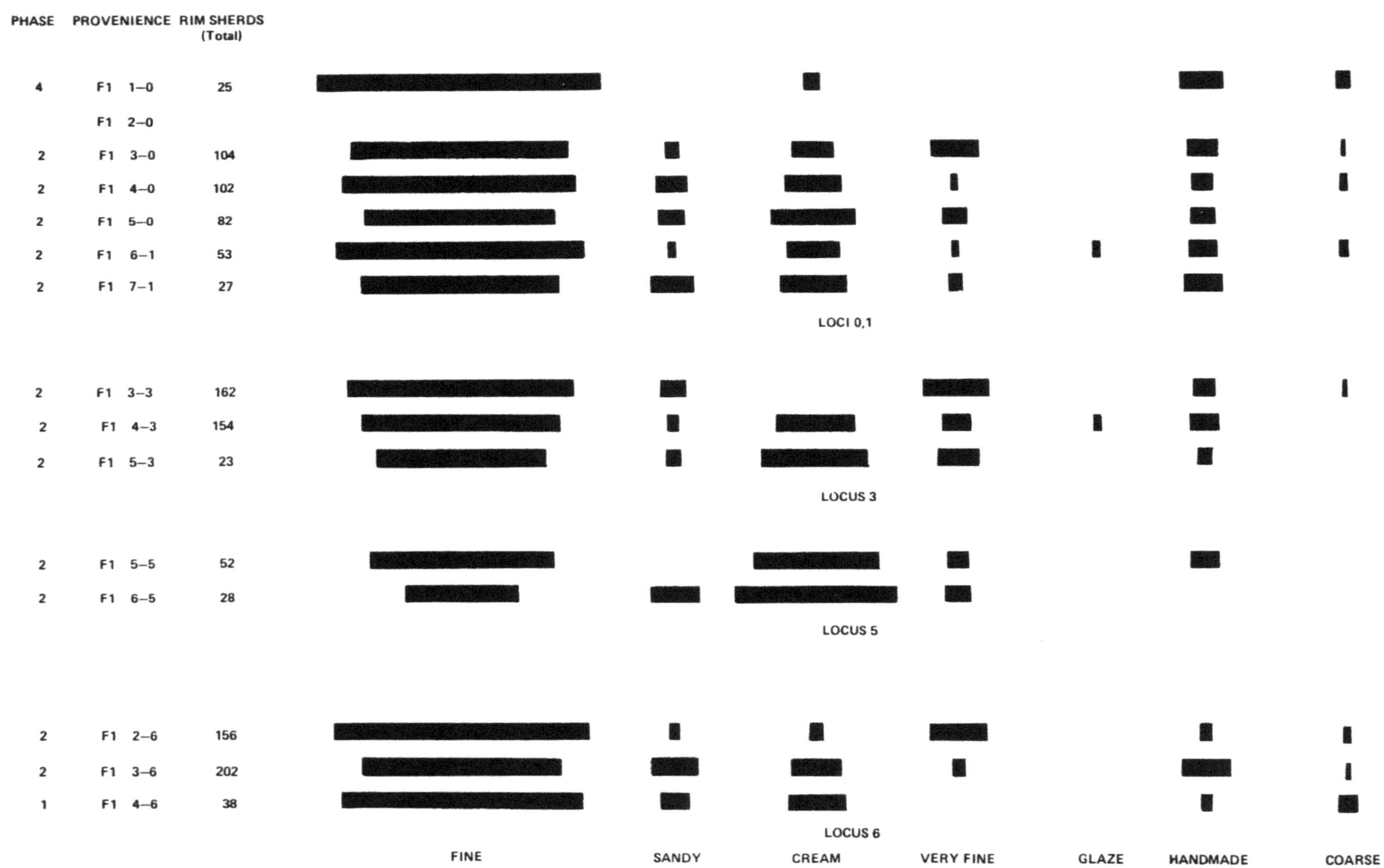


Figure 7.1. Bar diagram of pottery wares in F1 sounding. Scale: ▬ 10%.

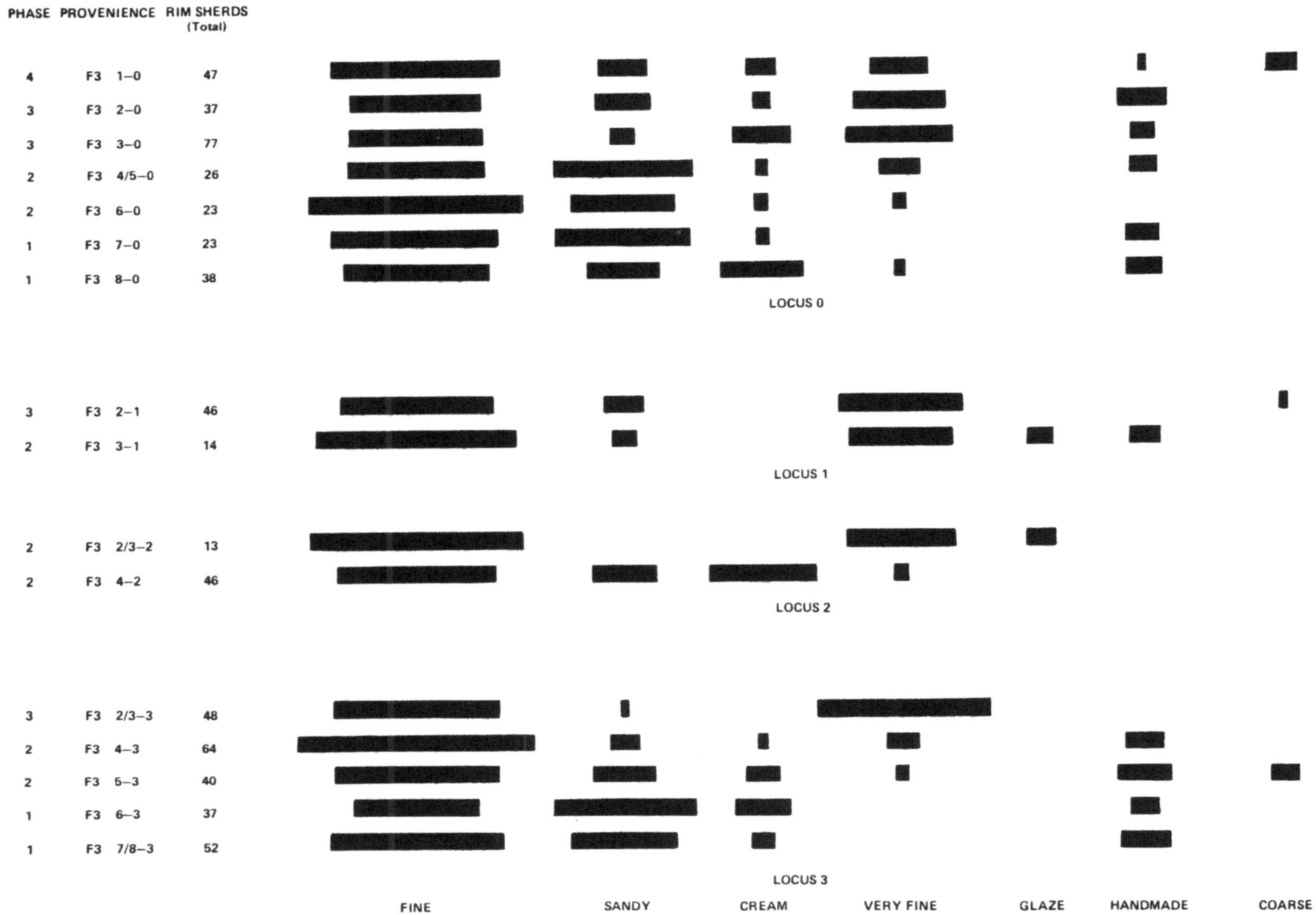


Figure 7.2. Bar diagram of pottery wares in F3 sounding. Scale: ▬▬ 10%.

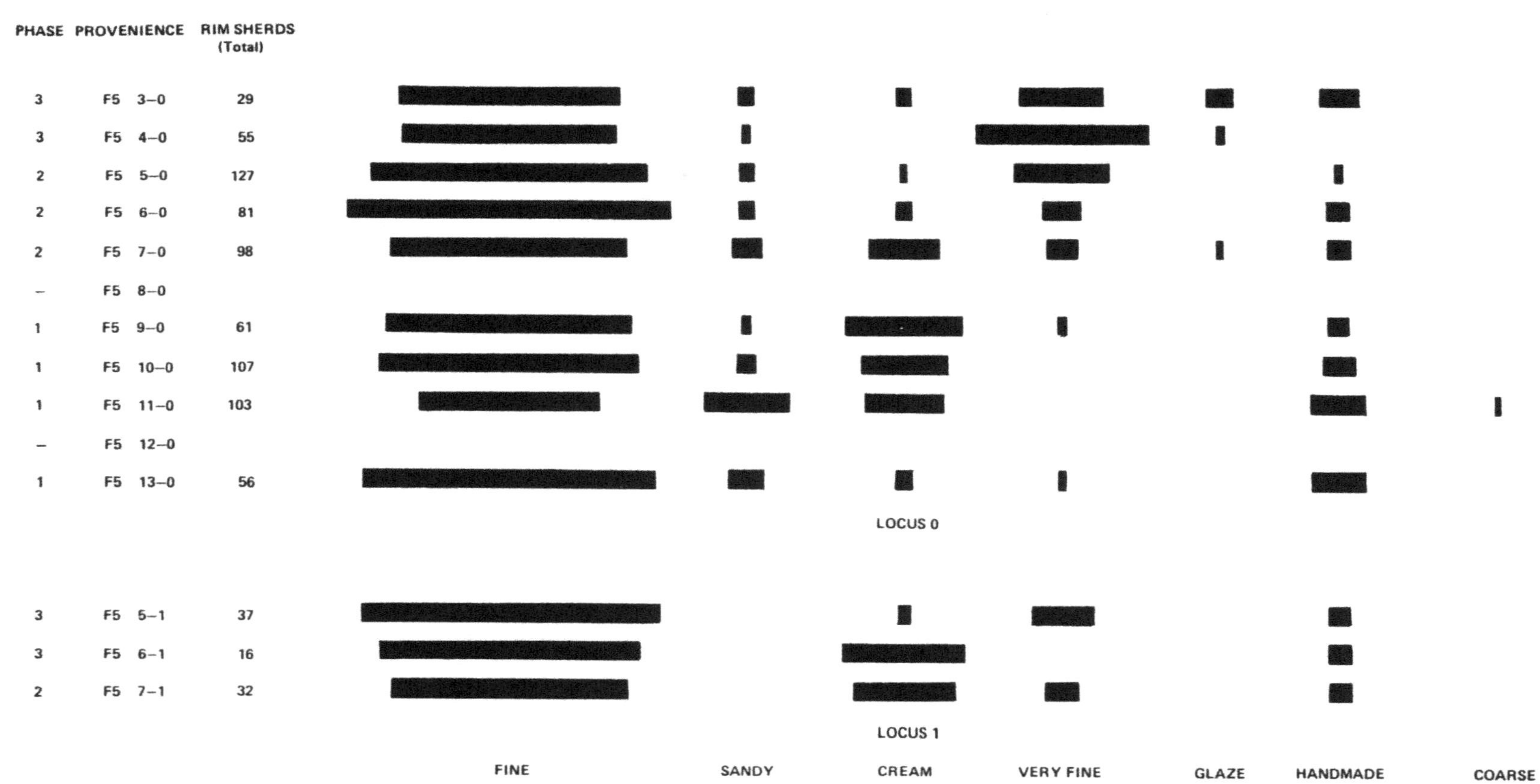


Figure 7.3. Bar diagram of pottery wares in F5 sounding. Scale: ▬ 10%.

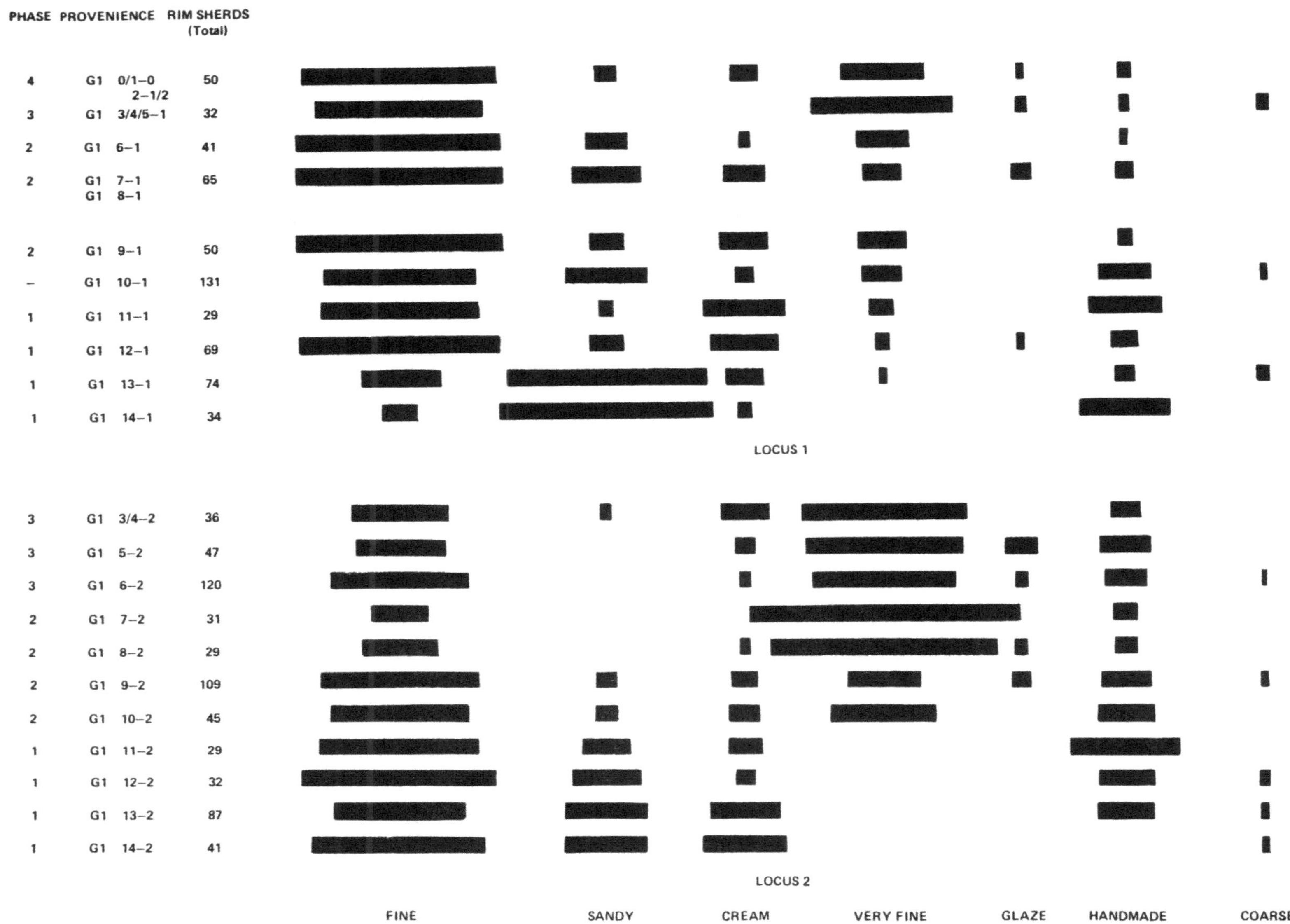


Figure 7.4. Bar diagram of pottery wares in G1 sounding. Scale: ▬ 10%.

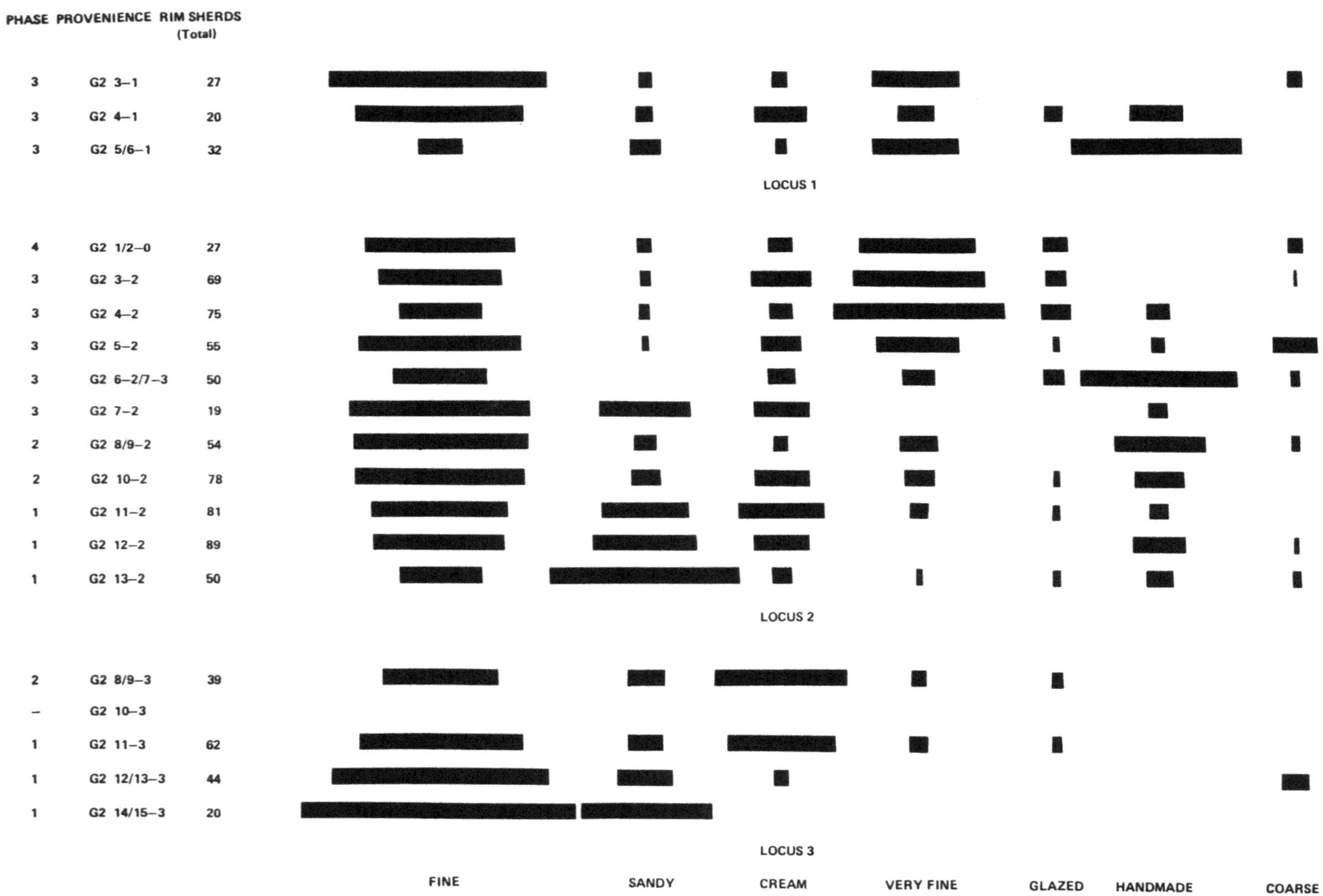


Figure 7.5. Bar diagram of pottery wares in G2 sounding. Scale: ▬ 10%.

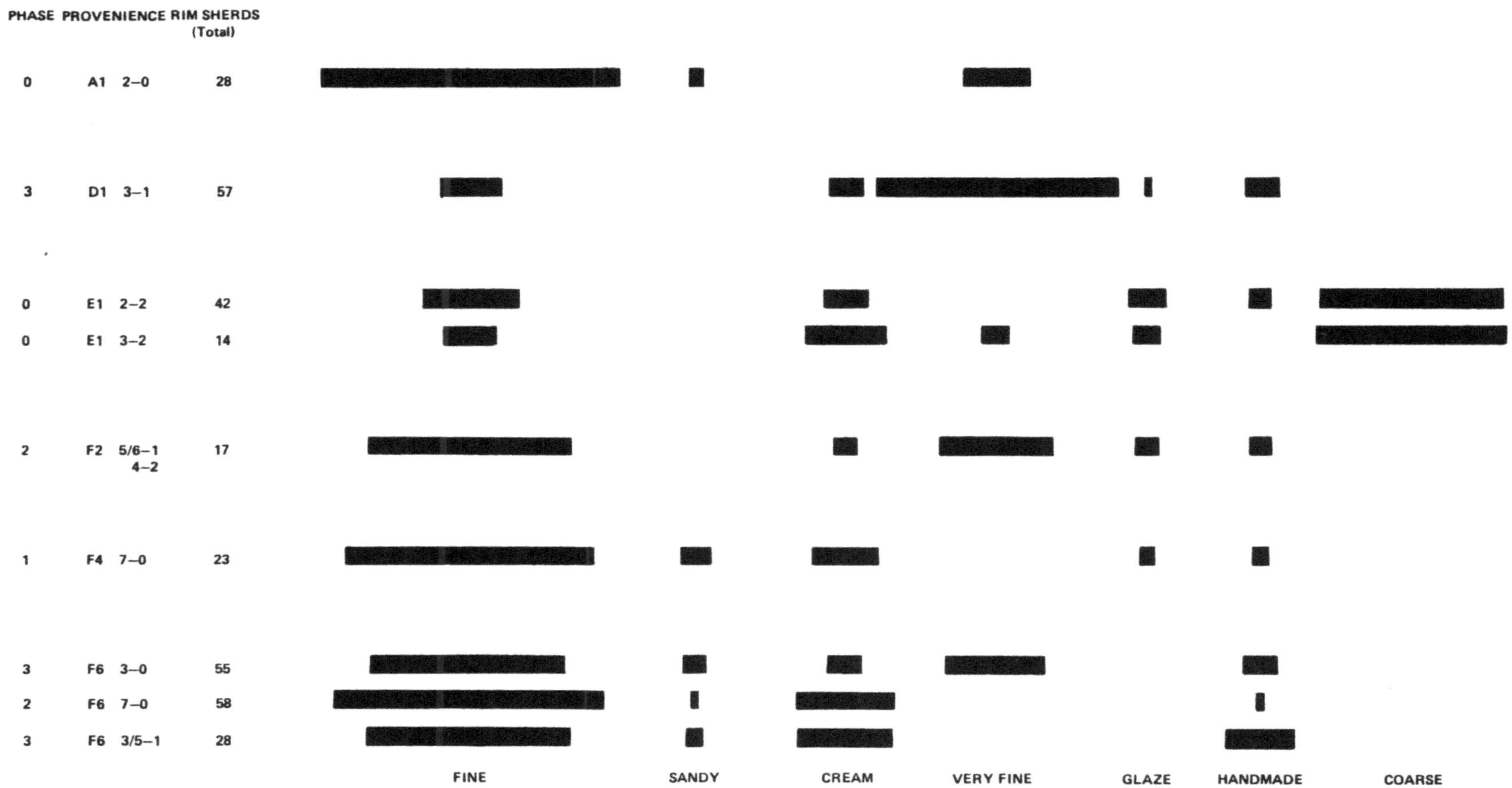


Figure 7.6. Bar diagram of pottery wares in A1, D1, E1, F2, F4, and F6 sounding. Scale: ▬ 10%.

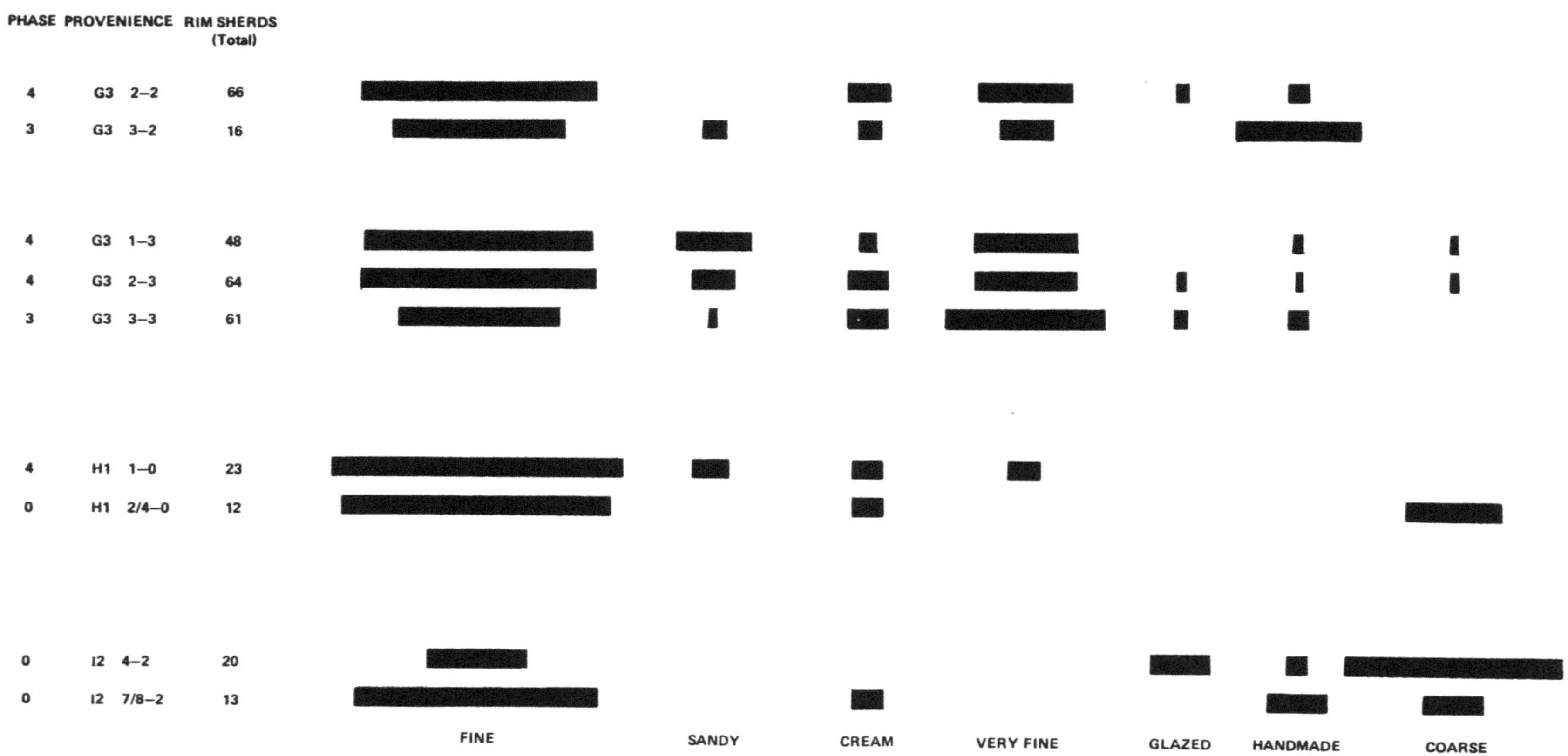


Figure 7.7. Bar diagram of pottery wares in G3, H1, and I2 soundings. Scale: ▬ 10%.

The distribution of sandy wares clearly shows that they were among the earliest wares produced at al-Basra and that, during the early phase, they were one of the most common wares in use. In addition, the relatively high concentration of sandy wares in deposits directly associated with the F3 kiln indicates that the structure may date to the early occupation phase at al-Basra. This probability is further supported by a thermoluminescence date for the kiln of A.D. 825 ±175 (Dur TL 34-4AS), around the time the city was established, although the high standard deviation is problemmatic (see Appendix 5). The predominance of sandy wares in the vicinity of the F3 kiln also suggests that they were fired in or near this structure. The sherds were found in fill deposits covering the kiln and may have originally come from a waster dump near the structure. A decline in the relative abundance of sandy wares in the second and third occupation phases (in some cases disappearing completely) shows that their production slowed down considerably. In the later assemblages, sandy wares, along with cream-colored wares, appear to be replaced by very fine buff wares, which were made from a different clay or by another clay preparation technique.

Cream-colored wares

Cream-colored wares are also relatively abundant in the early assemblages at al-Basra, but their distribution differs somewhat from that of sandy wares. As Figures 7.3-7.5 show, cream-colored wares reach their peak in levels that are slightly later in time than those in which sandy wares predominate. These levels are associated with architectural features that mark the collapse of the early city and usher in the major reconstruction of the center in the second phase. In addition, the proportion of cream-colored wares declines more gradually in the later phases than that of sandy wares. Cream-colored wares occur in moderately high proportions in the second occupation phase and in low proportions during the third occupation phase.

It is worth noting that cream-colored wares are relatively abundant in deposits associated with the F1 kiln (Figure 7.1), suggesting that they may have been produced in or near this structure. On the basis of the chronology, the F1 kiln may be dated to a slightly later time than the F3 kiln, sometime between the end of the first and the beginning of the second phase. It is possible that the production of cream-colored wares, which usually appeared in the form of a pitcher with a handle, was geared toward a Berber rather than an Arab market. This type of ware, which is occasionally painted with simple red or dark brown linear motifs, has been described as Berber ware in other parts of North Africa and Spain, both archeologically (Bel 1914; Zozoya 1969) and ethnographically (van Gennep 1911).

Very fine wares

Unlike sandy and cream-colored wares, very fine buff-firing wares are virtually absent from the earliest assemblages at al-Basra (see Figures 7.3-7.5). These finely textured sherds make their appearance in levels associated with architectural remains from the second occupation phase, where they are present in moderate quantities. They increase in relative abundance--sometimes very dramatically--in the final occupation phase. Because of their distribution, very fine wares serve as good diagnostic markers for middle and late phase assemblages. Their relatively high frequencies in the middle and upper levels of residential and industrial soundings and in shallow soundings located around the periphery of the site suggest that these areas were occupied during the later two phases. The relatively low proportions of very fine wares in deposits associated with the F1 kiln (Figure 7.1) suggests that the kiln was in operation during the middle occupation phase but that it was not involved in their production. By contrast, their nearly total absence from assemblages associated with the F3 kiln (Figure 7.2) provides further evidence for assigning this kiln to the early occupation phase at al-Basra. In any event, very fine wares are not closely associated with either kiln and were probably fired in a structure that has yet to be discovered at the site.

Glazed wares

Glazed wares are concentrated in deposits associated with the third and final occupation phase at al-Basra, although they also occur in earlier assemblages. The relative frequency of all glazed wares, which are considered as a group despite their diversity, amounts to far less than 5% in most provenience units. They occur in higher proportions in a few deposits associated with residential architecture, e.g., G1 and G2 (Figures 7.4-7.5), from the late occupation phase. They are also present in relatively high proportions in deposits associated with the city water cistern (E1) and the remains of an open-air lime kiln (I2). In these instances, the glazed material consists of a very thick, rather coarsely textured pottery that differs from the finer material found elsewhere on the site. Glazed wares are noticeably absent from the two kiln areas, F1 (Figure 7.1) and F3 (Figure 7.2), except in the upper levels.

Handmade, fine, and coarse wares

The relative abundances of the other three wares--fine buff-firing, handmade, and coarse--do not appear to increase or decrease in any consistent way from early to late levels. The proportion of fine buff wares, which predominate at the site, fluctuates from level to level, often in response to an increase or decrease in another ware category, but it does not show any unidirectional change. This ware appears to have been produced at al-Basra in fairly large quantities during all three occupation phases. The relative abundance of handmade wares, which served almost exclusively as cooking vessels, also does not vary consistently through time, although they seem to occur somewhat less frequently in the middle levels. The proportion of coarse wares is slightly higher in the early and late assemblages, but this trend does not occur in all of the soundings.

Summary of Ware Distribution

A summary of the relative abundances of each ware group by occupation phase are presented in Table 7.1. These percentages were calculated from the total number of sherds for each ware found in all provenience units assigned to a specific occupation phase. As the table illustrates, this data support the patterns that were observed in individual provenience units. Sandy buff wares drop from 23% in the

TABLE 7.1

COUNTS AND PERCENTAGES OF AL-BASRA WARES, BY OCCUPATION PHASE

| Occupation Phase | Total[1] | Fine | | Sandy | | Cream-Colored | | Very Fine | | Glazed | | Handmade | | Coarse | |
|---|---|---|---|---|---|---|---|---|---|---|---|---|---|---|---|
| | | (No.) | (%) | (No.) | (%) | (No.) | (%) | (No.) | (%) | (No.) | (%) | (No.) | (%) | (No.) | (%) |
| 4 Post-Urban | 350 | 205 | 58.6 | 29 | 8.3 | 29 | 8.3 | 60 | 17.1 | 8 | 2.3 | 11 | 3.1 | 8 | 2.3 |
| 3 Late | 1,109 | 442 | 39.9 | 36 | 3.2 | 86 | 7.8 | 388 | 35.0 | 29 | 2.6 | 116 | 10.5 | 13 | 1.2 |
| 2 Middle | 2,347 | 1,366 | 58.2 | 176 | 7.5 | 296 | 12.6 | 294 | 12.5 | 21 | 0.9 | 176 | 7.6 | 15 | 0.6 |
| 1 Early | 1,279 | 609 | 47.6 | 289 | 22.6 | 213 | 16.7 | 17 | 1.3 | 5 | 0.4 | 132 | 10.3 | 14 | 1.1 |

1 The total excludes 302 sherds that came from provenience units which could not be assigned to a phase.

early occupation phase to 3% in the third occupation phase. Cream-colored wares also decrease, from 17% to 8% during the same period. By contrast, very fine wares increase dramatically from less than 1% in Phase 1 to 35% in Phase 3. Glazed wares increase from about 1% in the early phase to 3% in the late phase. Fine buff wares, handmade wares, and coarse wares do not demonstrate any consistent pattern of change during the occupation of al-Basra.

## Temporal Distribution of Major Forms

The relative abundances of several vessel forms from al-Basra also vary through time. The most striking changes involve six of the 11 major forms: three types of small storage jars (triangular-rim jars [Form 41], band-rim jars [Form 42], and grooved-band-rim jars [Form 43]), carinated jars (Form 40), pitchers (Form 48), and nosed lamps (Form 93).

### Storage jars

As Figures 7.8-7.14 show, the relative frequencies of all three storage jars (Forms 41, 42, and 43) change in relation to one another over time. Band-rim storage jars (Form 42) are concentrated almost exclusively in assemblages associated with the earliest occupation levels, where they are present in moderate quantities; they are virtually absent in middle and late phase assemblages, a trend that is apparent in nearly all of the soundings. Grooved-band-rim jars (Form 43), likewise, are relatively abundant in early assemblages and gradually decline in relative frequency in later ones. By contrast, the proportion of triangular-rim jars (Form 41) steadily increases from early to late levels; in a few cases, the trend reverses itself briefly at points in the soundings where the deposits are likely to have been disturbed, e.g., in F1, G1, and G2 (Figures 7.8, 7.11, and 7.12) but overall it tends toward increasing frequency. In terms of storage vessels, then, band-rim and grooved-band-rim jars are prevalent in the early assemblages; band-rim jars disappear in the middle levels, with the other two types present in about equal proportions; in the late assemblages, triangular-rim jars become the dominant form. It is possible that the simultaneous changes in the proportions of these functionally similar but stylistically different storage jars reflects a shift in the number or size of the workshops that produced them.

### Pitchers

Pitchers (Form 48) exhibit an increase in relative frequency through time. The proportion of pitchers is relatively low in the early levels, it increases moderately in the middle levels, and it becomes the predominant vessel form in late assemblages. The relative abundance of pitchers rises steadily and somewhat steeply in the late levels, suggesting the possibility of a change in the eating or drinking habits of al-Basra residents.

### Carinated jars

The relative abundance of carinated serving jars (Form 40), which is moderately high in early assemblages, drops somewhat in the later

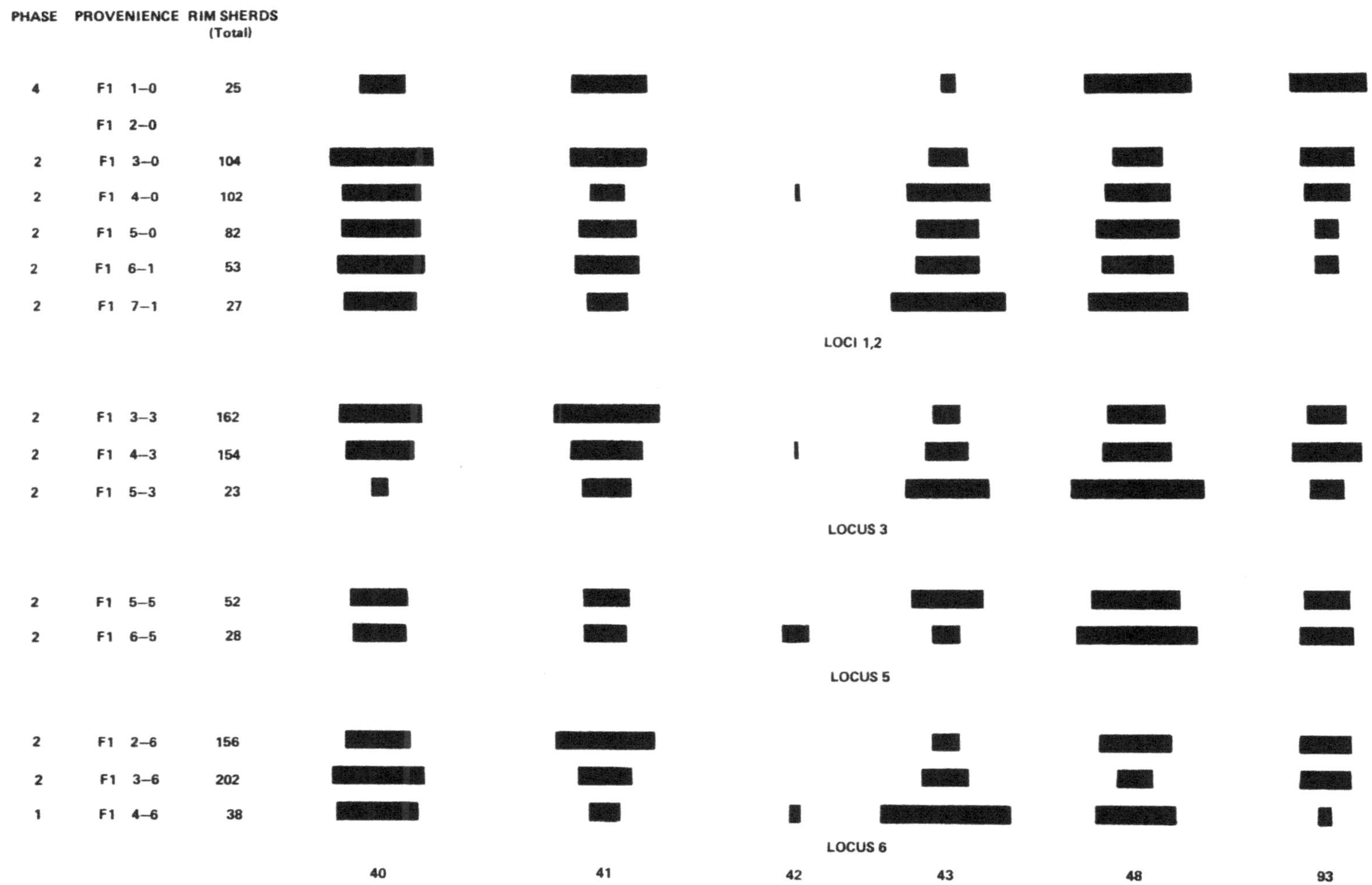


Figure 7.8. Bar diagram of major pottery forms in F1 sounding. Scale: ▬ 10%.

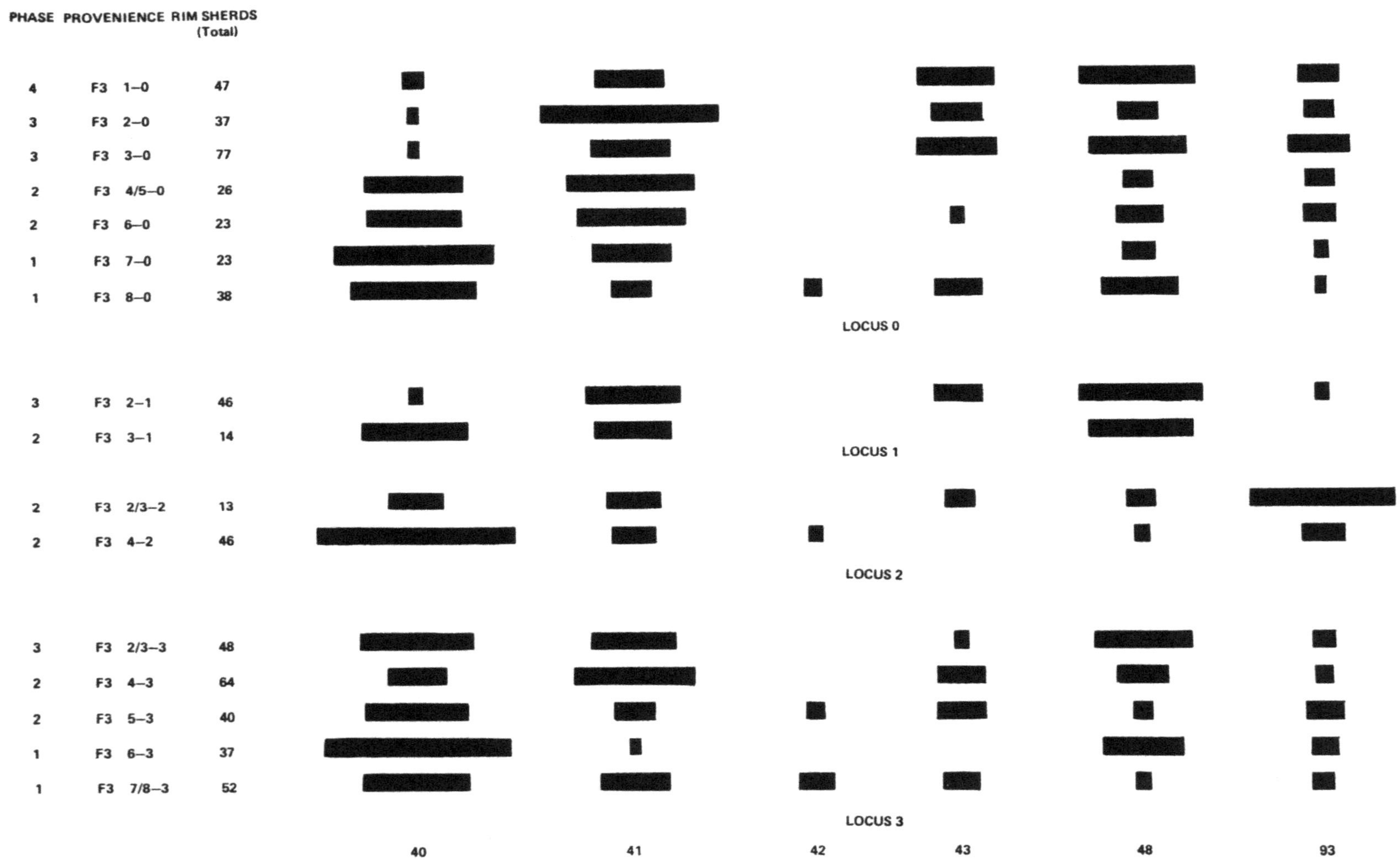


Figure 7.9. Bar diagram of major pottery forms in F3 sounding. Scale: ▬ 10%.

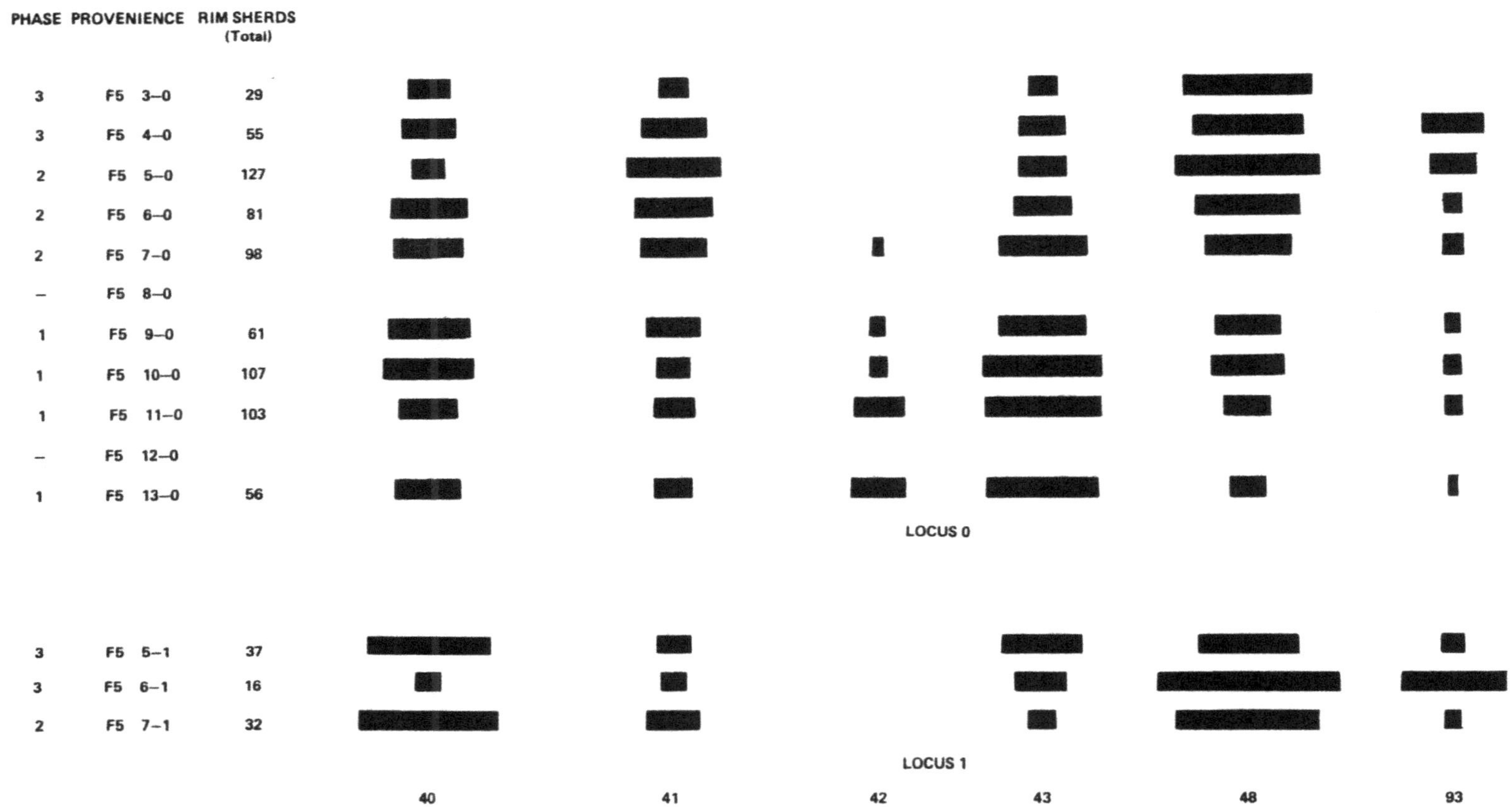


Figure 7.10. Bar diagram of major pottery forms in F5 sounding. Scale: ▬ 10%.

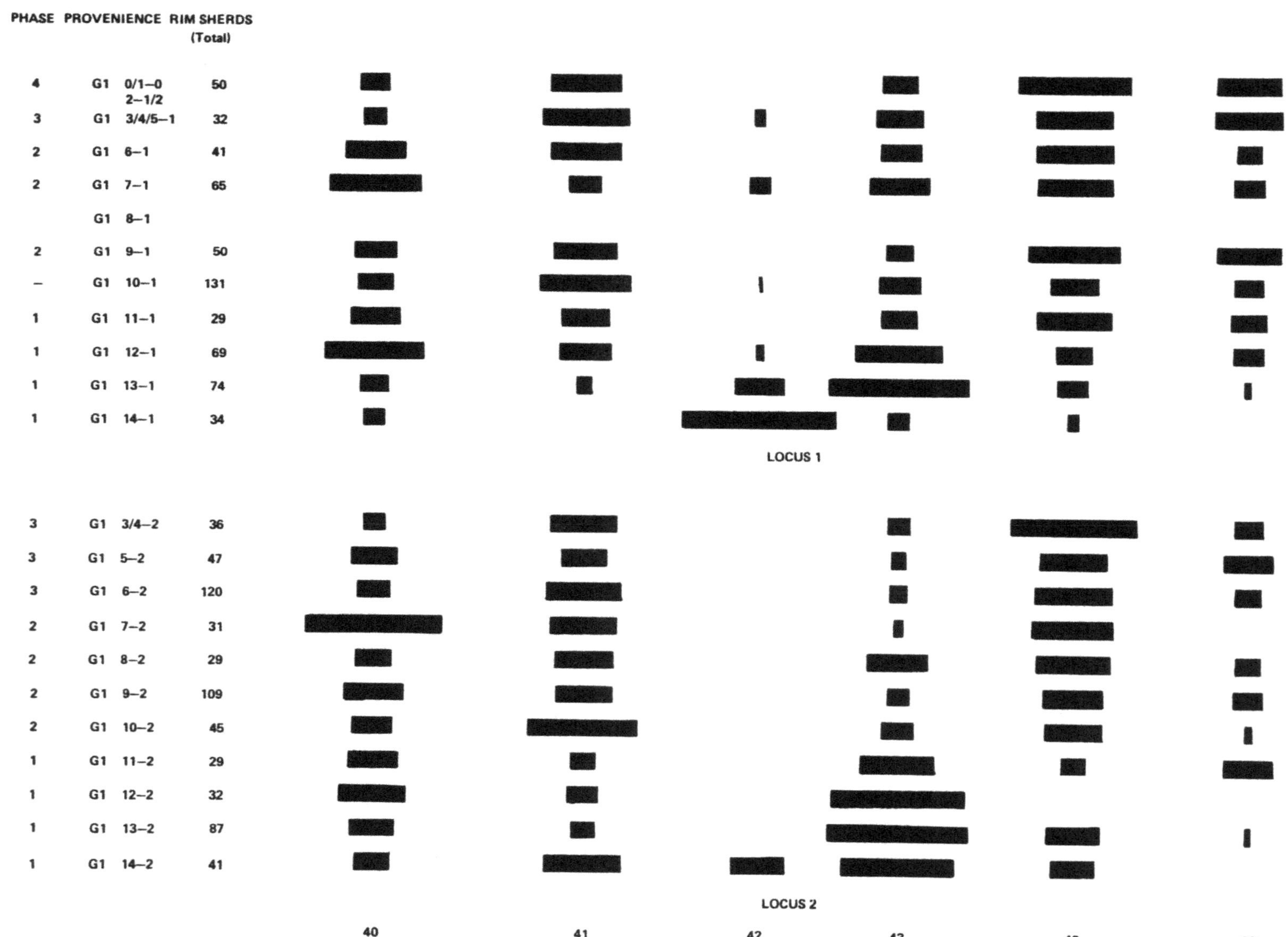


Figure 7.11. Bar diagram of major pottery forms in G1 sounding. Scale: ▬ 10%.

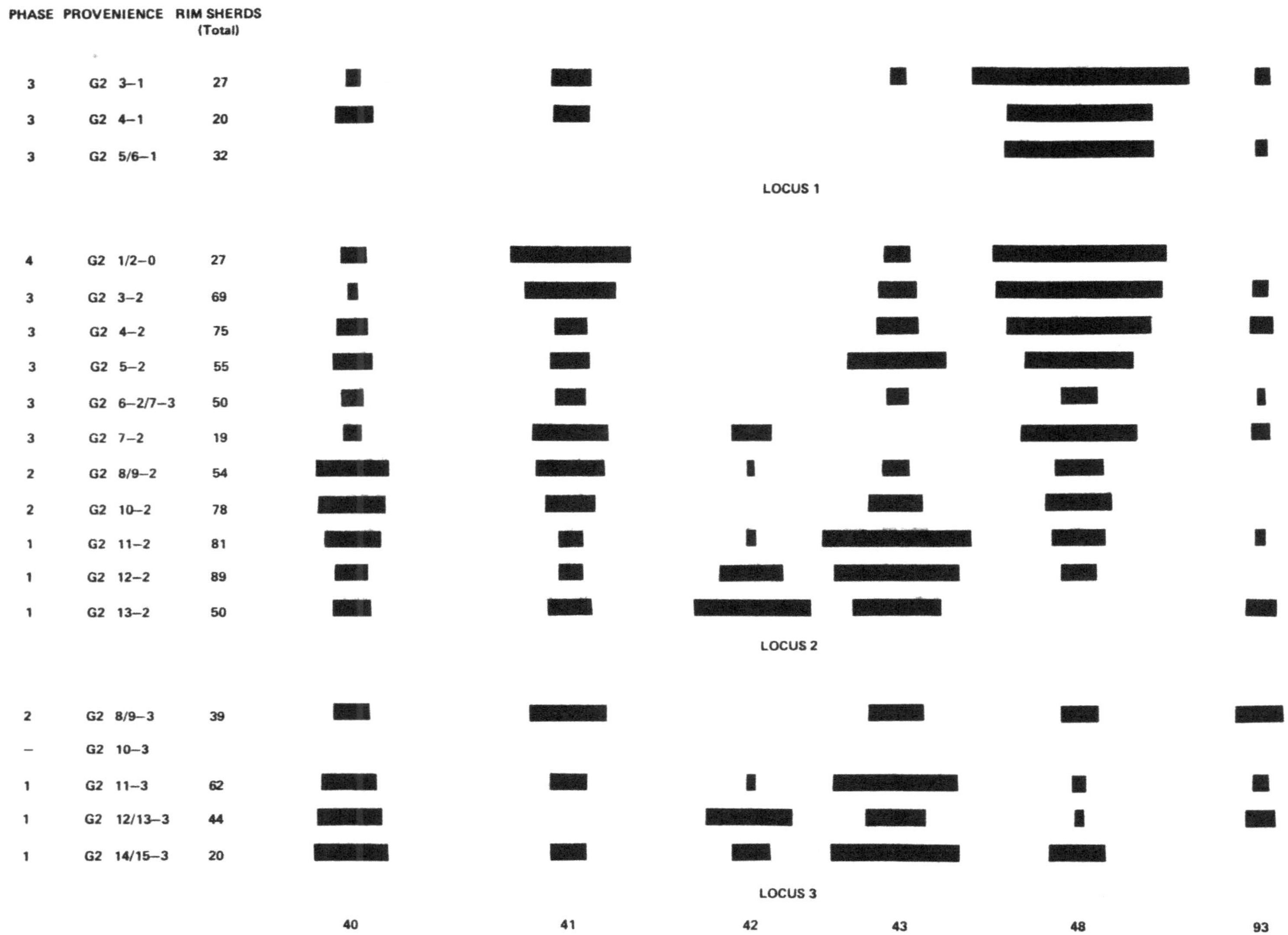

Figure 7.12. Bar diagram of major pottery forms in G2 sounding. Scale: ▬ 10%.

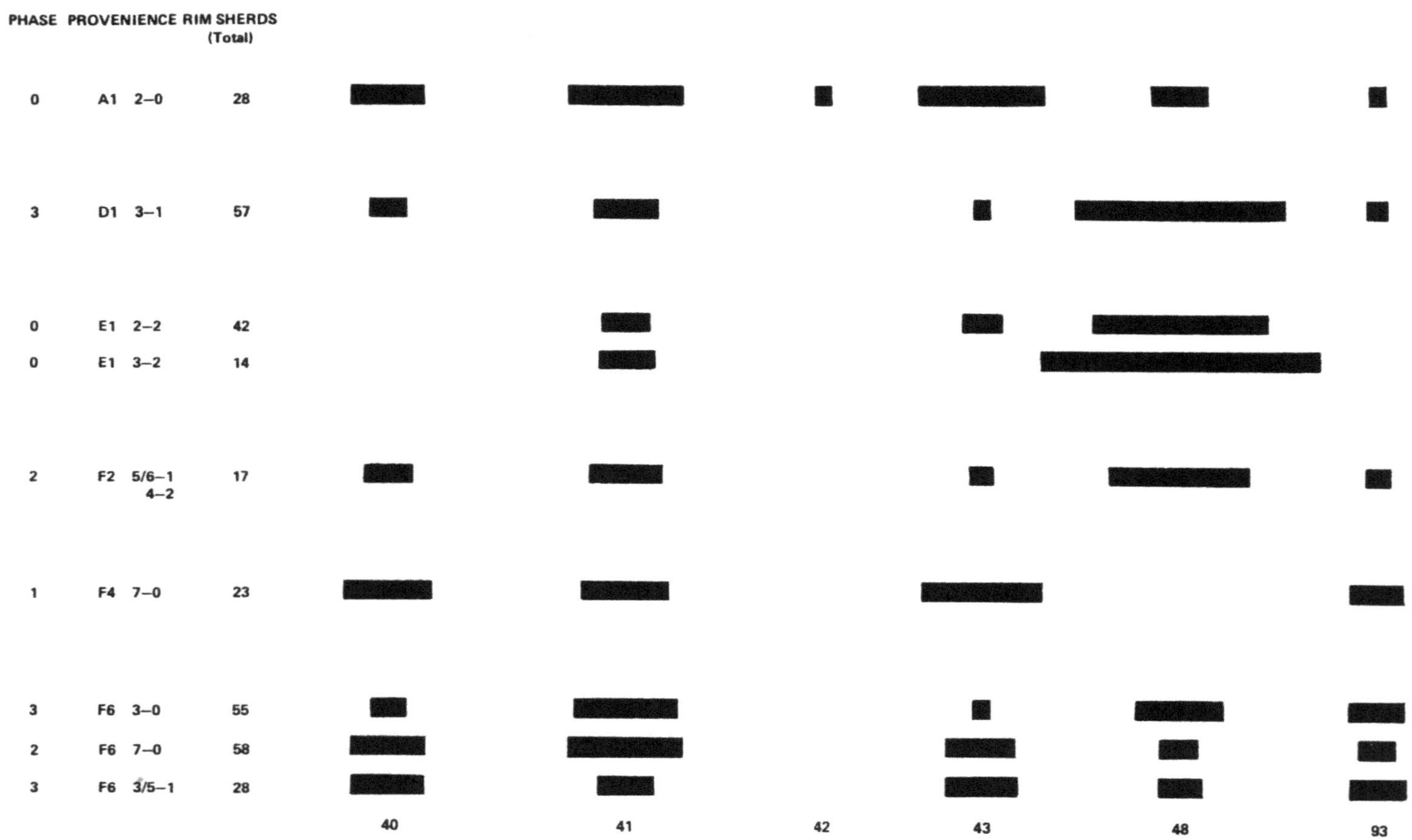


Figure 7.13. Bar diagram of major pottery forms in A1, D1, E1, F2, F4, and F6 soundings. Scale: ▬ 10%.

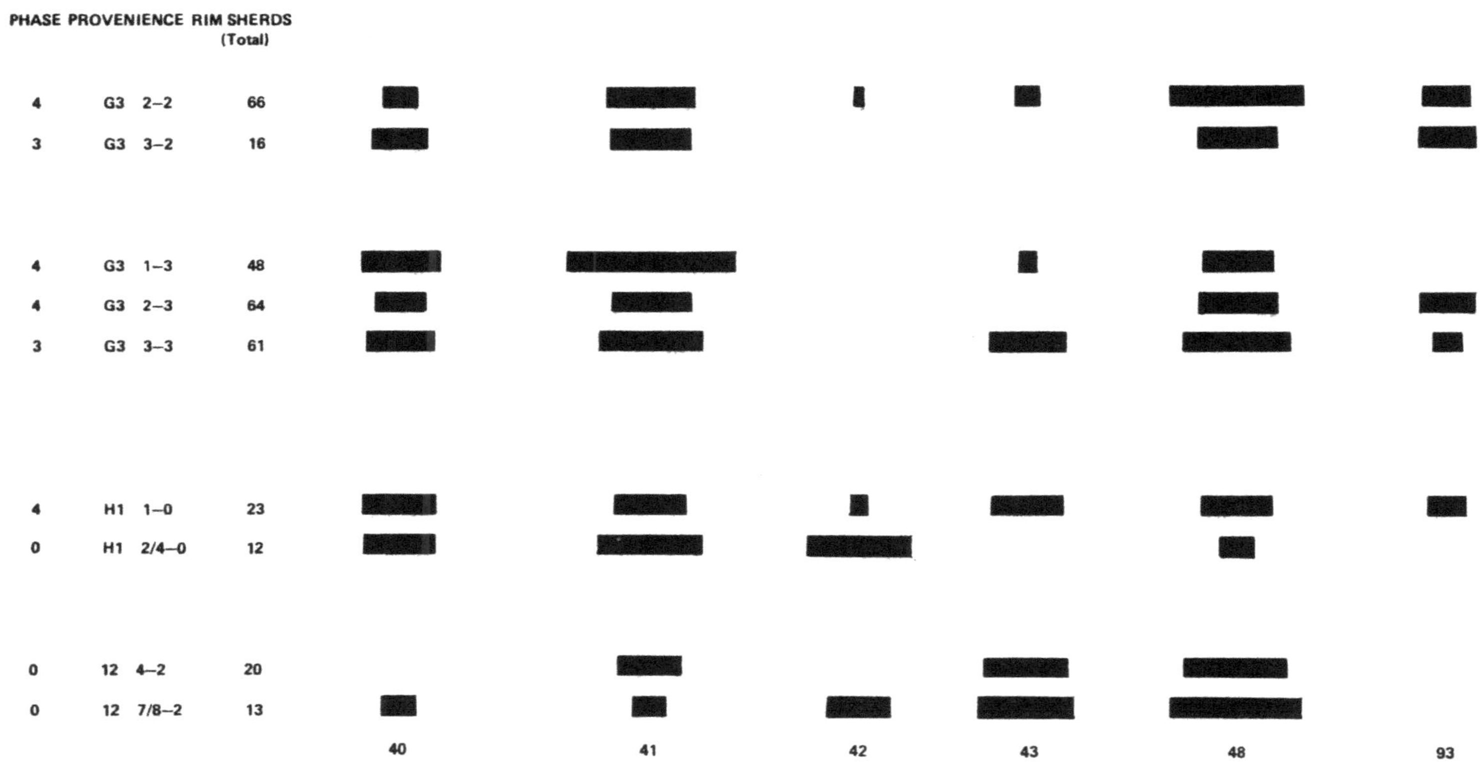


Figure 7.14. Bar diagram of major pottery forms in G3, H1, and I2 soundings. Scale: ▬ 10%.

deposits. This is particularly evident in the F3 and G2 soundings (Figures 7.9 and 7.12). It is also noteworthy that the relative frequency of this jar is much higher in deposits associated with the F3 kiln than in any other assemblages at the site, indicating that the vessel may have been fired in the kiln during the earliest phase of occupation at the site.

Nosed oil lamps

Nosed oil lamps (Form 93) also increase somewhat in relative frequency from early to late assemblages. The change is most clearly evident in the F1, F5, and G1 soundings (Figures 7.8, 7.10, and 7.11). The moderate increase in the proportion of lamp fragments may reflect a growing utilization of this form of lighting in the urban center, perhaps because of an overall rise in the standard of living.

Summary of Form Distribution

Changes in the relative abundances of forms at al-Basra by occupation phase are summarized in Table 7.2. These percentages were calculated in the same way as those in Table 7.1. The patterns described above are clearly evident in the summary. Three small storage jars (Forms 41, 42, and 43) change in relationship to one another. Form 42 drops from 9% in Phase 1 to less than 1% in Phases 2 and 3. Form 43 decreases from 27% in the first phase to less than 12% in the second and third phases. By contrast, Form 41 increases from 10% in the early phase to 18% in the last two phases. Overall, small storage jars decrease in proportion to the rest of the forms in the collection from about 45% in Phase 1 to about 28% in Phase 3. Among other forms, pitchers (Form 48) increase from 11% in the early occupation phase to 30% in the late phase. Lamps (Form 93) also increase a little, from 4% to 8% during the same period. Carinated jars (Form 40), on the other hand, drop from 18% to 10%.

Primarily because of their low frequencies, five other important forms failed to show a consistent change through time (Table 7.2). A tall-necked liquid storage jar (Form 44) varies between 3% and 5%. A large, wide-mouth storage jar (Form 74) changes slightly from about 2% in Phases 1 and 2 to 4% in Phase 3. The most common wheelmade bowl form (Form 12) remains at about 1% during the same time. Handmade cooking vessels (Forms 10 and 11) also fluctuate in relative frequency from early to late assemblages.

## Temporal Distribution of Minor Forms

The temporal distribution of minor vessel forms is presented in Table 7.3. These 26 forms, which encompass a variety of bowls, jars, flasks, lamps, cups, and basins, together constitute about 5% of the al-Basra pottery collection. Because of their low frequencies, most of the forms do not show any clear pattern of change. A few of them, however, seem to be present exclusively in one or another assemblage. Three jar forms (76, 78, 79) and one flask form (71), for example, appear exclusively in the first occupation phase. Another jar form (50) and flask form (69) are present only in Phase 1 and 2 assemblages. A bowl form (14), jar form (55), and lamp form (94) appear exclusively

TABLE 7.2

COUNTS AND PERCENTAGES OF MAJOR AL-BASRA FORMS, BY OCCUPATION PHASE

| Occupation Phase | Rim Total[1] | Form 40 | | Form 41 | | Form 42 | | Form 43 | | Form 48 | | Form 93 | |
|---|---|---|---|---|---|---|---|---|---|---|---|---|---|
| | (No.) | (No.) | (%) | (No.) | (%) | (No.) | (%) | (No.) | (%) | (No.) | (%) | (No.) | (%) |
| 4 Post-Urban | 350 | 36 | 10.3 | 97 | 27.7 | 2 | 0.6 | 34 | 9.7 | 98 | 28.0 | 31 | 8.9 |
| 3 Late | 1,110 | 110 | 9.9 | 202 | 18.2 | 1 | 0.1 | 112 | 10.1 | 331 | 29.8 | 87 | 7.8 |
| 2 Middle | 2,347 | 468 | 19.9 | 442 | 18.8 | 14 | 0.6 | 282 | 12.0 | 431 | 18.4 | 230 | 9.8 |
| 1 Early | 1,279 | 231 | 18.1 | 122 | 9.5 | 111 | 8.7 | 345 | 27.0 | 142 | 11.1 | 52 | 4.1 |

| Occupation Phase | Form 44 | | Form 74 | | Form 10 | | Form 11 | | Form 12 | | All Major Forms | |
|---|---|---|---|---|---|---|---|---|---|---|---|---|
| | (No.) | (%) | (No.) | (%) | (No.) | (%) | (No.) | (%) | (No.) | (%) | (No.) | (%) |
| 4 Post-Urban | 16 | 4.6 | 2 | 0.6 | 9 | 2.6 | 1 | 0.3 | 5 | 1.4 | 331 | 94.6 |
| 3 Late | 48 | 4.3 | 44 | 4.0 | 83 | 7.5 | 21 | 1.9 | 18 | 1.6 | 1,057 | 95.3 |
| 2 Middle | 112 | 4.8 | 42 | 1.8 | 139 | 5.9 | 27 | 1.2 | 37 | 1.6 | 2,224 | 94.8 |
| 1 Early | 41 | 3.2 | 18 | 1.4 | 84 | 6.6 | 44 | 3.4 | 14 | 1.1 | 1,204 | 94.1 |

1 Rim total excludes 302 sherds from proveniences that could not be assigned to an occupation phase.

in Phase 2 deposits. The recovery of additional pottery will help to determine if, indeed, these minor forms are reliable markers for various temporal phases at al-Basra.

TABLE 7.3

COUNTS OF MINOR AL-BASRA FORMS, BY OCCUPATION PHASE

| Occupation Phase | Rim Total[1] (No.) | Bowls 13 | 14 | 15 | 16 | 17 | Basins 19 | 20 | 25 |
|---|---|---|---|---|---|---|---|---|---|
| 4 Post-Urban | 350 | 1 | 0 | 1 | 0 | 3 | 0 | 1 | 0 |
| 3 Late | 1,110 | 2 | 0 | 1 | 3 | 0 | 0 | 1 | 3 |
| 2 Middle | 2,347 | 26 | 2 | 1 | 3 | 0 | 0 | 3 | 13 |
| 1 Early | 1,279 | 5 | 0 | 1 | 0 | 0 | 0 | 0 | 4 |

| Occupation Phase | Miscellaneous Jars 49 | 50 | 52 | 53 | 54 | 55 | 56 | 72 | 76 | 78 | 79 |
|---|---|---|---|---|---|---|---|---|---|---|---|
| 4 Post-Urban | 0 | 0 | 2 | 4 | 0 | 0 | 0 | 6 | 0 | 0 | 0 |
| 3 Late | 1 | 0 | 5 | 1 | 5 | | 2 | 8 | 0 | 0 | 0 |
| 2 Middle | 8 | 3 | 4 | 7 | 7 | 1 | 5 | 12 | 0 | 0 | 0 |
| 1 Early | 3 | 7 | 6 | 13 | 9 | 0 | 2 | 6 | 1 | 5 | 2 |

| Occupation Phase | Flasks 68 | 69 | 70 | 71 | Lamps 94 | Cups 110 | Minor Form Total[2] (No.) | (%) |
|---|---|---|---|---|---|---|---|---|
| 4 Post-Urban | 1 | 0 | 0 | 0 | 0 | 0 | 19 | 5.4 |
| 3 Late | 7 | 0 | 1 | 0 | 0 | 12 | 52 | 4.7 |
| 2 Middle | 12 | 3 | 0 | 0 | 3 | 9 | 122 | 5.2 |
| 1 Early | 4 | 4 | 2 | 1 | 0 | 1 | 76 | 5.9 |

[1] The total excludes 302 sherds from provenience units that could not be assigned to an occupation phase.
[2] The total excludes 26 sherds from provenience units that could not be assigned to an occupation phase.

## Temporal Distribution of Glazed and Painted Pottery

In addition to looking at the proportions of al-Basra wares and forms, I examined the relative abundances of glazed pottery and painted wares. This information provides further evidence for temporal changes in the al-Basra pottery collection. It also sets the stage for a comparison of the al-Basra material with other Islamic pottery collections from North Africa and Spain, most of which have been studied from the point of view of decorative techniques.

As we saw in Chapter 5, glazed wares constitute a very small proportion of the al-Basra collection. Despite their small numbers, they encompass a range of glaze colors and decorative treatments. As Table 7.4 shows, many of them change in relative abundance from one occupation phase to the next. The most common type--an undecorated green and/or honey-colored ware--varies from about 50% to 67% of all glazed material. Three other types occur almost exclusively in the earliest occupation phase. One of these is a light yellow (bisque) glazed ware decorated with a green-and-manganese motif. Another type is incised with a simple linear motif under a glaze. A third type is a mottled light green, possibly tin-based, glazed ware. Another type--a glaze-painted ware--occurs in small amounts in all three occupation phases. This ware is a possible precursor of _cuerda seca_, a type of glazed pottery that appeared in the western Mediterranean in the mid-tenth century. Finally, four types of glazed pottery are present only in the last two occupation phases. Two of them have a honey- or green-colored glaze over a white slip. A variant has a honey-colored glaze over green paint; this one occurs mostly in the third phase. A fourth is the distinctive manganese decorated, honey-colored glazed ware, which is found at many later Islamic sites in Morocco; it occurs almost exclusively in assemblages associated with the final occupation phase at al-Basra.

Painted pottery is also relatively rare in the al-Basra collection. Unlike the variety of glazed types, however, painted wares are relatively homogeneous in terms of decorative technique and design. They are almost invariably decorated with simple linear and dot motifs in a red- or brown-matte paint. They usually belong to pitcher-like vessels (Form 48) with a cream-colored paste, similar to the wares produced by modern-day Berbers. As Table 7.5 shows, painted sherds are concentrated in the early occupation levels at the site. The relative abundance of painted sherds in the total al-Basra collection drops from 4% in Phase 1 to about 1% in Phases 2 and 3.

## Summary of al-Basra Pottery Chronology

The al-Basra pottery sequence can be divided into three major phases, each associated with specific architectural features at the urban site. These phases are characterized by the appearance or the predominance of certain pottery wares and forms.

Phase 1, which is associated with architecture from the earliest settlement at the site, is characterized by a predominance of sandy buff-firing and cream-colored wares, some of which are painted, and by

TABLE 7.4

PERCENTAGES OF AL-BASRA GLAZED POTTERY, BY OCCUPATION PHASE

| | | Undecorated | | | | | | | | | Decorated | | | | |
|---|---|---|---|---|---|---|---|---|---|---|---|---|---|---|---|
| Occupation Phase | Glazed Total[1] (No.) | Green (%) | Honey (%) | Honey/ Green (%) | Brown (%) | Yellow (%) | Light Green (%) | Honey-on White (%) | Green-on White (%) | Honey-on Green (%) | Manganese-on Honey (%) | Green Manganese (%) | Glaze Painted (%) | Cuerda Seca (%) | Incised (%) |
| 4 Post-Urban | 89 | 32.6 | 20.2 | 1.1 | 21.4 | 3.4 | 1.1 | 5.6 | 2.3 | 1.1 | 6.7 | 0.0 | 3.4 | 0.0 | 1.1 |
| 3 Late | 143 | 35.0 | 10.5 | 4.2 | 7.0 | 1.4 | 1.4 | 7.7 | 4.2 | 11.2 | 11.2 | 0.0 | 4.2 | 1.4 | 0.7 |
| 2 Middle | 61 | 34.4 | 24.6 | 8.2 | 3.3 | 1.6 | 0.0 | 9.8 | 4.9 | 1.6 | 1.6 | 0.0 | 6.6 | 0.0 | 3.3 |
| 1 Early | 32 | 25.0 | 18.8 | 12.5 | 6.3 | 0.0 | 6.3 | 0.0 | 0.0 | 3.1 | 0.0 | 9.4 | 6.3 | 0.0 | 12.5 |

[1] Glazed total includes all rim, base, handle, and body sherds. The total excludes 17 sherds to an occupation phase.

TABLE 7.5

COUNTS OF AL-BASRA PAINTED POTTERY, BY OCCUPATION PHASE

| Occupation Phase | Rim Total[1] (No.) | Painted Rims (No.) | (%) |
|---|---|---|---|
| 4 Post-Urban | 350 | 3 | 0.9 |
| 3 Late | 1,109 | 13 | 1.2 |
| 2 Middle | 2,347 | 19 | 0.8 |
| 1 Early | 1,279 | 45 | 3.5 |

[1]Rim total excludes 302 sherds from provenience units that could not be assigned to an occupation phase.

the absence of very fine wares. It is also marked by the presence of band-rim storage jars (Form 42) and a high proportion of grooved-band-rim storage jars (Form 43) relative to triangular-rim jars (Form 41). Green-and-manganese-on-light yellow glazed sherds are also present in these early assemblages.

Phase 2, which is related to a major rebuilding of the city involving the construction of well-shaped limestone walls and plaster floors, is distinguished by the appearance of very fine buff wares and a decline in the proportion of sandy buff and cream-colored wares. It is also associated with the nearly complete disappearance of band-rim jars (Form 42) and a low proportion of grooved-band-rim jars (Form 43) relative to triangular-rim jars (Form 41), which become the dominant form of small storage jars.

Phase 3, which is associated with a second reconstruction of the city involving roughly built stone walls, is characterized by a relatively high proportion of very fine buff wares and glazed wares, particularly a manganese-painted, honey-colored variety. It is also characterized by a high proportion of pitchers (Form 48) and triangular-rim jars (Form 41) and a low proportion of grooved-band-rim jars (Form 43) and carinated jars (Form 40).

Phase 4, which is associated with ephemeral settlements that took place after the city's final collapse, is characterized by an abrupt reversal in many of the trends described above. The assemblages from this post-urban phase contain residual pottery that has been eroded from the surface of the site during the last centuries.

The changes evident in the al-Basra ceramic assemblages are the result of a number of different, but closely interrelated cultural phenomena. The shift from sandy to very fine buff-firing wares in the middle occupation phases, for example, represents a technological innovation on the part of some al-Basra potters. The impetus for this change may have come from inside the local pottery community, with the discovery of new clay sources or clay treatment methods, or from new potters who migrated to al-Basra, bringing with them new technological knowledge and skills. Because they closely resemble other wares made at al-Basra in terms of vessel form and paste composition, very fine wares were likely developed by local potters as an improvement over the coarser sandy buff-firing wares. The increase in their relative abundance in the third occupation phase indicates that they were accepted and ultimately preferred by al-Basra residents, possibly at a time when the standard of living in the city was on the upswing. It is possible that the introduction of very fine wares at al-Basra stimulated a certain amount of competition among local potters, perhaps marking the demise--or the retooling--of workshops that produced sandy wares.

The change in the proportion of cream-firing wares, on the other hand, more likely represents a shift in the ethnic character of the urban center. As we noted earlier, cream wares seem to have been produced by Berber potters for a Berber market. If this were true, the concentration of cream-firing wares in the earliest levels of the site would show that the Berbers constituted a significant part of the original population at al-Basra. The decrease in the proportion of cream-firing wares in later assemblages would indicate that the Berber population declined or that it had been assimilated into Arab culture, although it did not disappear entirely.

The small but distinct increase in imported glazed wares at al-Basra reflects another cultural phenomena--a shift in regional trade or exchange networks. The increase in glazed pottery, particularly manganese-on-honey wares, in the late assemblages suggests that the merchants and/or residents of the city gained access to the products of a new manufacturing center. This access may have come about because of the lifting of previous political and economic barriers, for example, the opening of southern Spain to Morocco after the fall of the Idrisid dynasty and the withdrawal of the Fatimids from the Maghreb.

Cultural processes, such as technological innovation, economic competition, and outside commercial ties, seem to be implicated to some extent in the chronological changes evident in the al-Basra pottery assemblages. As we will see in Chapter 8, these phenomena need to be distinguished from one another in any study that deals with ceramic diversity and standardization.

## Survey of Early Islamic Pottery in North Africa and Spain

Relatively little is known about early Islamic pottery in the western Mediterranean. Until recently, few sites or components of sites dating from the eighth to the eleventh century had been investigated at all (see Table 7.6). Most of the sites that have been excavated, particularly those in the western Maghreb, date to a later period or contain important architectural remains, such as those at Ksar es-Seghir (Redman 1985), Belyounesh (Grenier 1980, Terrasse 1976), Ceuta (Posac Mon 1960, 1962), and Marrakesh (Deverdun and Rouche 1949, Terrasse and Meunier 1957). In addition, much of the pottery that has

TABLE 7.6

EARLY ISLAMIC SITES IN MOROCCO, ALGERIA, TUNISIA, AND SPAIN

| Date | Morocco | Algeria | Tunisia | Spain |
|---|---|---|---|---|
| 700 | | | | |
| | | | | Umayyad Emirate |
| | Nakur | | | (711-942) |
| | | | | Cordoba |
| | Idrisids | | | |
| 800 | (789-985) | Idrisids | Aghlabids | |
| | Fez | (790-973) | (800-909) | |
| | al-Basra | Tlemcen | Kayrawan | |
| | Asila | | 'Abbasiyya | |
| | | | Rakkada | |
| 900 | | | | |
| | Fatimids | | Fatimids | |
| | (917-985) | | (909-975) | Umayyad Caliphate |
| | | | Mahdiyya | (942-1035) |
| | | | Sabra Mansuriyya | Medina az-Zahra |
| | Zenatas | | | Medina Elvira |
| 1000 | (985-1062) | Hammadides-Zirides | | |
| | | (1007-1080) | Zirides | Taifas |
| | | Qal'a | | (1035-1087) |
| | Almoravids | Bougie | | |
| | (1062-1140) | | | Almoravids |
| | Marrakesh | Almoravids | | (1087-1180) |
| 1100 | Salé | (1080-1145) | | |
| | Dchira | | | |
| | Almohads | Almohads | Almohads | |
| | (1140-1243) | (1145-1264) | (1160-1264) | Almohads |
| | | | | (1180-1232) |
| 1200 | | | | |

been recovered from these sites has been poorly provenienced, because of a lack of good stratigraphic controls during excavations (Golvin 1980; Mohamedi 1980). Furthermore, most of the material that has been retained consists of highly decorated pottery or other museum quality material (Vitelli 1981). Finally, much of the Islamic pottery has not been published, including some recently excavated material from 'Abbasiyya, Sabra Mansuriyya, and Rakkada in Tunisia (Daoulatli 1980).

With the exception of al-Basra, no other early Islamic site in Morocco has been investigated in a systematic manner. Only a small part of the site of Nakur, which was built by the Arabs in the eighth century, was excavated in 1980 before it was buried under dam construction on the Nakur River (Redman 1983, n.d.). The Islamic levels at Fez and Taza, the major entrepôts established by the Idrisids in the eighth and ninth centuries, have been examined only sporadically (Ricard and Delphy 1931). The Islamic components at Tangier, Azila and Kurt, among other sites, have not been examined in any substantial way. Sigilmasa, an important trade entrepôt in the Sahara, has been scheduled for excavation only in recent years. Although some ceramic material from this early period in Morocco's history has found its way into museum collections, much of it has remained unprovenienced or unpublished.

The situation is somewhat different in Algeria and Tunisia where more archeological work has been focused on early Islamic sites (i.e., from the Aghlabid, Fatimid, and Hammadid periods). In the central Maghreb, the city of Tlemcen (formerly Agadir), established by the Idrisids in the ninth century, was excavated in the early part of this century (Bel 1914). The eleventh-century Hammadid sites of Qal'a and Bougie have been reexamined more recently (Golvin 1965, 1980; Marcais 1913, 1916). The Islamic components of classical sites, such as Sétif, are currently under excavation (Février et al. 1970, Mohamedi 1980). In Tunisia, the major commercial center of Kayrawan, founded by the Aghlabids in the eighth century, and their royal palaces at 'Abbasiyya and Rakkada, were investigated in the early part of this century (Marcais 1928) and were reexamined recently, but the newly recovered material has not yet been published. Likewise, the tenth-century Fatimid capital of Mahdiyya and royal palace at Sabra Mansuriyya (Marcais 1924; Zbiss 1956) and the Fatimid-Hammadid components of Carthage (Ferron and Pinard 1954) were examined in the early part of the century and have been reinvestigated in the last few years. Only the re-analysis of the Carthage Islamic pottery has been published (Vitelli 1981). A useful summary of the Islamic material from many of these sites also has been published recently, although it is limited to decorative wares, particularly those that resemble Syrian and Iraqi Islamic wares (Daoulatli 1979, 1980).

In Spain, too, a number of important Islamic sites from the Umayyad period have been excavated, among them the Caliphal palaces at Medina az-Zahra near Cordoba and Medina Elvira (Ilbira) near Malaga (Velaquez Bosco 1912, 1923). A relatively comprehensive synthesis of early Islamic pottery from southern Spain, based primarily on an examination of museum collections, was published recently (Zozoya 1980a, 1980b), making the Spanish Islamic material among the best documented in the western Mediterranean.

Despite its limitations, the ceramic data from other early Islamic sites in the Maghreb and Spain lets us place the al-Basra pottery into a broader temporal and regional context. The al-Basra pottery shows some similarities to contemporaneous material from Négrine and 'Abbasiyya in Tunisia, and Medina az-Zahra, Medina Elvira, and other sites in Spain, all of which date to the ninth and tenth centuries. The al-Basra collection, however, differs in several respects from pottery assemblages excavated at Salé and Dchira in Morocco and at Tlemcen, Qal'a, and Bougie in Algeria, most of which date to the eleventh and twelfth centuries. The al-Basra collection also diverges in several ways from the excavated Nakur material, which seems to derive from eleventh- or twelfth-century contexts at the earliest. The comparative data allow us to assign dates--albeit tentatively--to ceramic material from specific occupation phases at al-Basra.

## Eighth- to Eleventh-Century Islamic Pottery

### North Africa

The pottery excavated from Nakur appears to date to the eleventh or twelfth century rather than to the eighth century when the city was founded. The material comes mostly from two soundings excavated to a depth of 1.5 m to bedrock on a hilltop west of the Nakur River (Redman 1983). Like that from al-Basra, the material consists largely of unglazed wares, a few of which are decorated with incised lines around the vessel neck or shoulder (Redman 1983:Figure 8a, h) and very few glazed pieces. By contrast, the Nakur material contains a much larger proportion of handmade wares than the al-Basra material (30% compared with 5% at al-Basra). Some of the Nakur forms are similar to those found at al-Basra, such as a spoutless pitcher (Redman 1983:Figure 8g, h, k), a tall-necked jar with a snub-nosed rim (Redman 1983:Figure 8i), and a glazed bowl (Redman 1983:Figure 8m, p); also the bases of most vessels are flat, or slightly convex, like those at al-Basra. On the other hand, several Nakur forms are entirely unknown at al-Basra, including a three-legged jar with a double carinated body (Redman 1983: Figure 8o), a green-glazed, brown-and-white painted jar with a filter in the neck (Redman 1983:Figure 8t, u), and a pedestaled lamp, all of which came from the lowest levels excavated at the site. As we will see later, these forms clearly belong to a genre of vessels that appear in the eleventh and twelfth centuries in the Maghreb. On this basis, we can suggest that the Nakur pottery sample probably dates to *no earlier* than the eleventh century. It is very possible that earlier pottery exists at the site but it was not recovered during the brief excavations conducted there.

The pottery from the Tunisian sites of 'Abbasiyya and Négrine date fairly clearly to the ninth and tenth centuries. The palace at 'Abbasiyya, a few kilometers south of Kayrawan, was occupied by the Aghlabids from 801 to 877 and then settled by others until the late 900s (Marcais 1928). The oasis town of Négrine was settled sometime in the eighth or ninth century but was definitely abandoned by the middle of the eleventh century when the Hilalian Arabs invaded Tunisia (Pianel 1951). The pottery from these sites consists of both glazed and unglazed wares. Unglazed wares are generally decorated with incised lines, both horizontal and undulating, and with finger- and tool-impressed designs on strips of clay applied to vessel surfaces (Pianel 1951:Plates 3-4). Glazed pottery is decorated almost invariably with brown (manganese), green, or blue motifs on a variety of different glazed bases. The most common styles are manganese motifs on a green or white tin-based glaze; manganese and/or green motifs on a white tin-based or yellowish glaze; and more rarely manganese and blue motifs on a white glaze (Marcais 1928:Figures 15-17; Pianel 1951:Plates 7-21). According to Daoulatli (1980), green glazed motifs on a yellowish base were particularly popular during the Aghlabid period. One fine example noted by Marcais (1928:Figure 16) features a delicate green floral design on a bisque-colored base; the sinuous stems are surmounted by pointed leaves and large circles. This motif is similar to that found on two green-on-light-yellow (bisque) specimens excavated from the earliest levels at al-Basra (see Figure 5.3d). The motifs on glazed pottery from the Tunisian sites usually consist of geometric, floral, animal, or epigraphic themes. A variation of a glazing technique that Marcais calls _céramique à réserve_ also occurs in the early assemblages; this technique consists of applying an opaque glaze directly on a clay surface to form a design; although Marcais does not say how the glaze was prevented from running, it is possible that a substance like wax was used. A few examples of this technique occur at al-Basra, in a small group described as glaze painted wares. The decorative techniques and styles of Tunisian glazed wares are believed to have come from pottery-making centers in Iraq and Syria, either directly by importing the pottery or indirectly through the artisans trained in the techniques (Marcais 1928; Pianel 1951). It is significant that these early Islamic collections do not appear to include any examples of stamp-impressed or _sgraffito_ pottery, which usually are present in eleventh- or twelfth-century contexts in the western Mediterranean.

Spain

According to Zozoya (1980a, 1980b), the early Islamic Spanish pottery can be attributed to at least five different periods: the Umayyad Emirate (756-942); the Umayyad Caliphate (942-1035); the period of the Taifas (1035-1085); the Almoravid period (1085-1180); and the Almohad period (1180-1232). The pottery from the first two periods, which is most important to the al-Basra study, is marked by an abrupt change in decorative techniques and vessel forms around A.D. 940. According to Zozoya, the change occurred when large numbers of artisans migrated from the eastern Mediterranean to build and furnish the new administrative-residential palaces at Medina az-Zahra and Medina Elvira for the recently established Caliphate. A gradual change in the third period coincides with a decline in the power and influence of the royal court. The appearance of new elements in the last two periods is attributed to the invasion of two powerful Berber groups from North Africa--the Almoravids and Almohads.

During the Umayyad Emirate, Spanish pottery retains traces of Roman and Visigothic influences, as manifested in the form of large, glazed carinated bowls with stamped impressions, small jars with slightly convex bases and decorated with white paint, and pitchers with tri-cornered mouths (Zozoya 1980a:Figures 1a-b, 2a-b, 3a-b), none of which are found in the al-Basra collection. The period is also marked by the appearance of red- and brown-painted wares, which are usually decorated with simple vertical stripes or linear motifs (Zozoya 1969, 1980b). According to Zozoya, the ware was introduced into Spain by North African Berbers, possibly during the reign of Abd al-Rahman II between A.D. 822 and 852 (Zozoya 1980b:312). This genre of painted pottery occurs at al-Basra, particularly in assemblages associated with the early occupation phase. A distinctive feature of Emirate pottery is the prevalence of flat or slightly convex bases on all common wares and on the green-and-manganese-glazed wares that appear toward the end of the period. As we mentioned earlier, flat bases are also predominant at al-Basra.

At the beginning of the Umayyad Caliphate in 942, Spanish pottery undergoes an abrupt change, particularly at Medina az-Zahra and Medina Elvira where the Caliphal court undoubtedly demanded high quality luxury goods. According to Zozoya, this period marks the introduction of imported lusterware and Chinese porcelain from the eastern Mediterranean and _cuerda seca_, whose origin is unknown. It also marks the apogee of ring-based, green-and-manganese-decorated white tin-glazed plates, which seem to have been made at both palace complexes. A variety of monochrome and bichrome wares, including manganese-decorated, honey-colored pottery, are also present at this time. Unglazed wares are usually incised or, less frequently, painted with red, black, or white linear motifs or decorated with stamped designs. Only a few of these decorative techniques, including monochrome and bichrome glazing, _cuerda seca_, and incising and painting, are present at al-Basra; green-and-manganese-on-white glazed sherds appear to be totally absent. In terms of forms, the Spanish material from this period includes a wide variety of bowls, plates, jars, small jars, jugs, bottles, cups, cooking pots, and nosed lamps. A number of the bowls, cooking pots, and lamps (Zozoya 1980a:Figures 5, 10, 11) resemble forms commonly found at al-Basra, but others do not.

Other important changes in Spanish Islamic pottery occur during the next three periods. The Taifas period is characterized by the appearance of a decorative technique called _verdugones_, a variant of _cuerda seca_ in which the vessel surface is only partly glazed (Zozoya 1980a: 281). In addition, convex bases are almost entirely replaced by ring bases (Zozoya 1980a:Figures 13, 14, 15) and filters, some with floral designs, begin to appear in the necks of two-handled jars. The Almoravid period is distinguished by the introduction of _sgraffito_, a decorative technique that involves scratching a design through a slip or glaze; this technique is believed to have originated in northern Africa (Zozoya 1980b:315). The Almohad era is marked by the continuing presence of _sgraffito_ and by the disappearance of filters from jars.

The similarities in decorative techniques and vessel forms attest to a certain amount of interaction between al-Basra and southern Spain. It is likely that this interchange occurred among the general populace, rather than among the elite. With the exception of _cuerda seca_, the decorative techniques and vessel shapes common to both places are peculiar to utilitarian wares rather than to high-status, decorative pottery. For example, the al-Basra collection contains examples of red- or brown-painted pottery associated with Berbers and flat bases on utilitarian wares but no examples of the fancy green-and-manganese-painted plates with ring bases manufactured in the Caliphal court or the lusterware and Chinese porcelain imported from the Middle East.

The analogs also provide a temporal framework for the al-Basra pottery. The presence of red- and brown-painted pottery in assemblages associated with the original settlement of al-Basra, historically dated to the early 800s, coincides with the appearance of Berber wares in Spain, estimated by Zozoya to have occurred sometime between 822 and 852. The presence of this pottery in the earliest levels at al-Basra confirms that the site was indeed of Islamic foundation, not Roman, as some authors have suggested (Tissot 1877; Eustache 1955), and that the site was probably settled in the late 700s or early 800s.

The appearance of *cuerda seca* pottery, manganese-on-honey glazed wares, and ring bases in assemblages associated with the third and final occupation phase at al-Basra allows us to date these assemblages tentatively to some time *after* 940. As we saw above, these ceramic features make their debut in Spain with the establishment of the Caliphate. The fact that they make their appearance at al-Basra at about the same time supports the idea that they may have come from Spain. In addition, their occurrence at al-Basra suggests that the Moroccan city may have become more oriented toward Spain in the mid- to late-tenth century. This would corroborate the documentary evidence that indicates that Spain extended its political influence into northern Morocco in the late tenth century after the Fatimids withdrew from the Maghreb.

Finally, the total absence at al-Basra of *sgraffito* wares, which appear in Spain with the Almoravid invasion in about A.D. 1085, suggests that al-Basra may have been abandoned by that time.

## Eleventh- and Twelfth-Century Islamic Pottery

### North Africa

The divergence in decorative techniques and vessel shapes between al-Basra and several eleventh- and twelfth-century sites in Morocco and Algeria provides further evidence for a *terminus ante quem* for al-Basra of about A.D. 1050. The pottery from Tlemcen, which Bel assigned to two groups--the earlier one to the late tenth- early eleventh centuries and the later one to the eleventh-twelfth centuries (Bel 1914), is remarkably similar in many respects to that from al-Basra. The bulk of the Tlemcen material consists of unglazed and undecorated wheelmade wares, along with a small number of handmade vessels. A smaller portion of the collection was decorated with horizontal or undulating incised lines or bands of lines around the vessel neck or shoulder. Some of it was painted with red or brown linear or geometric motifs. The glazed pottery from the earlier assemblage was not decorated, although the later material was, including green and brown (manganese) decorated tin-glazed bowls and plates and a form of *cuerda seca*, in which yellow and green designs are outlined in brown, possibly like the examples from late assemblages at al-Basra (see Figure 5.3e, f). Several Tlemcen vessel forms are also similar to those found at al-Basra, including triangular-shaped jar rims, flat or slightly convex bases, and nosed lamps (Bel 1914:Figures 15, 17, 20, 21). Unlike the al-Basra material, however, the Tlemcen collection also contains numerous examples of stamped pottery. The pottery was decorated with stamped motifs placed around the edges of plates and serving bowls; some of it was covered with a white slip or green glaze, although most of it was plain. Stamped pottery seems to have been made at Tlemcen, as evidenced by the discovery of several stamps in the excavations.

The pottery from Qal'a of the Banu Hammad is dated fairly reliably to the beginning of the eleventh century when the mountaintop site was settled by the Hammadids (ca. 1010-1090) at the end of the twelfth century (Marcais 1913; Golvin 1965, 1980). Like assemblages at al-Basra and Tlemcen, the major part of the collection consists of unglazed wares that were usually decorated with incised lines (Golvin 1965:197). Unlike the al-Basra material, however, the rest of the Qala pottery consists primarily of glazed wares, most of which had brown, green, yellow, or violet designs on a white, tin-glazed background. There were also some stamped and sigillé wares, either plain or covered with green glaze, and, more rarely, lusterware and cuerda seca. In terms of vessel forms, the Qal'a material also diverges from that of al-Basra. In general, the vessels have raised bases; nosed lamps have no funnel and a short handle or a handle with a small conical protuberance; and large storage jars, which stand up to 80 cm in height, are often decorated with floral or geometric motifs around their necks (Golvin 1980:Figures 2, 3, 6, 7). In addition, entirely different forms are present, including footed braziers and cylindrical vessels (Golvin 1980:Figures 5, 8). Thus, aside from the decoration found on utilitarian wares, the material from Qal'a differs substantially from that at al-Basra in terms of decorative techniques and vessel forms.

The differences are even more remarkable in the ceramic material from Salé, Dchîra, and Taza, all of which are closer geographically to al-Basra than any of the other sites described above. The pottery from Salé dates to no earlier than the eleventh century (Delphy 1951-1952, 1955), and that from Taza (Ricard and Delphy 1931) and the Almohad fortress at Dchîra to the twelfth century (Terrasse 1937). Stamped pottery is relatively abundant at all three sites. In fact, a number of stamps were found at Salé, as at Tlemcen, indicating that the pottery was very likely produced there. The stamps usually consist of floral or epigraphic motifs (e.g., Ricard and Delphy 1931:Plates 7-24). Cuerda seca pottery is also present in these collections, particularly at Salé, where white, green, brilliant yellow, and dark brown geometric designs cover entire vessel surfaces (Delphy 1951-52:Plate 5). Other glazed material from these sites includes manganese-on-honey glazed wares, manganese-and-green decorated, white tin-glazed vessels, and a few pieces of lusterware. There are also some examples of sgraffito at Salé. The vessel forms are also strikingly different from those at al-Basra. Raised bases, including ringed, footed, and flanged varieties, are predominant. Pedestaled lamps, cylindrical vessels, and ceramic well coverings (margelle de puit), which were often elaborately decorated with stamped impressions (Delphy 1955:Plates 3-9; Ricard and Delphy 1931:Plates 25-30), were also present.

The ceramic material from these eleventh- and twelfth-century sites provides further evidence for a terminal date of no later than the eleventh century for al-Basra. The pottery from these eleventh- and twelfth-century sites is characterized by stamped decoration and a variety of new forms, such as raised bases, cylindrical vessels, footed braziers, and pedestal lamps--none of which occur at al-Basra.

## Summary and Conclusions

Using both the ceramic and historical evidence, we can assign the al-Basra pottery collection to a period of time that begins around A.D. 800, when the city presumably was founded, and ends sometime in the eleventh century, probably before or during the Almoravid invasion of Morocco in A.D. 1070. In addition, we can tentatively suggest that the al-Basra pottery *on the whole* represents a provincial industry that may have evolved from an indigenous pottery tradition with its roots in the late Roman period. The vessel forms that dominate the al-Basra assemblage--carinated jars (Form 40) and small and large storage jars with short, curved necks (Forms 41, 42, 43 and 74)--show stronger resemblances to late Roman forms (Hayes 1972) than to the bowls and plates that seem to occur at most Islamic sites. This remains speculative because the shapes of utilitarian vessels from other early Islamic sites in North Africa are not well known. We can say with some degree of confidence, however, that the al-Basra collection contains pottery with Berber influence, specifically the red- and brown-painted pitchers with one or two handles that are most prevalent in the early phase. The imported glazed wares that appear in the al-Basra collection indicate that the urban center was in contact with other cities in the Maghreb and Spain from the time of its original settlement in about A.D. 800. We can also suggest that the glazed wares present in the earliest al-Basra assemblages exhibit some similarities to the Aghlabid-Fatimid material in Tunisia, in particular the green and/or manganese-on-light yellow glazed pottery. In addition, the suite of glazed wares that appears in late assemblages at al-Basra shows a closer resemblance to Spanish Umayyad pottery, specifically manganese-on-honey glazed and *cuerda seca* wares. Again, this remains hypothetical because of the gap in our knowledge about the origin and diffusion of many early Islamic glazing techniques in the western Mediterranean. It may be possible to resolve some of these questions in the future by undertaking provenience studies of glazed wares from sites that are chronologically well defined, such as 'Abbasiyya, Négrine, Sabra Mansuriyya, Medina az-Zahra and Medina Elvira, as well as al-Basra.

# CHAPTER 8

# CHANGES IN THE ORGANIZATION OF POTTERY PRODUCTION AT AL-BASRA

One of the main objectives of this research has been to reconstruct the organization of the pottery industry at al-Basra and, as far as possible, to identify the factors that were involved in its change through time. In many ways, the al-Basra collection constitutes a good case study because of the availability of pottery from well-defined occupation phases, the unambiguous evidence of pottery production at the site, and the availability of historical information about the site itself and about Islamic craft production in general. This study brings this evidence together in an effort to develop a detailed picture of ceramic production at the urban center during the early Islamic period, a time of rapid political and economic change in the Maghreb. The broad questions that it addresses are: (1) to what extent did political or administrative factors have an impact on the pottery industry centered at al-Basra, (2) to what degree did economic factors stimulate changes in its organization and technology and, (3) is it possible to distinguish between these factors in the archeological record.

While a number of different indicators have been used by other researchers to investigate changes in the organization of ceramic production (e.g., Feinman et al 1981; Hagstrum 1985), this study focuses on one: the diversity, or variety, of pottery assemblages. As it is used here, the concept of diversity refers both to the number of different classes of items present in an assemblage (the dimension of *richness*) and to the uniformity with which items are distributed among the classes (the dimension of *evenness*). This chapter presents the rationale for the use of the diversity concept in ceramic analysis. It applies a diversity index to the al-Basra pottery in order to infer the relative number and size of artisan workshops during each occupation period and, in a rough way, to monitor any changes in the degree of competition among potters. The chapter then discusses these results in terms of the Feinman-Rice scenario, in which political and administrative factors are implicated in the reorganization of a ceramic industry, and in light of the entrepreneurial model, in which economic forces are considered primarily responsible for changes in production organization.

## Diversity Concept in Ceramic Analysis

One of the most sensitive indicators of the organization of craft production is the diversity, or variety, of locally made pottery. In an ethnographic study of North African potters, Balfet (1965) found that the diversity of pottery produced in domestic contexts varied markedly from that made by semi-skilled and skilled artisans, in terms of both technological and decorative characteristics. She attributed this to differences in the relationships between potters and their pottery and potters and the users of the pots in each of these contexts. Balfet found that, in terms of technique and decorative styles, domestically produced pottery was spatially heterogeneous but temporally homogeneous, that is, it varied greatly from place to place within a region but it did not change very much through time. On the other hand, she found that the techniques and decorative styles of artisan-produced pottery were spatially homogeneous but temporally heterogeneous, that is, they were relatively uniform within a region but they were subject to innovation and change with the passage of time.

In a similar way, ceramic diversity can be used to infer changes within a predominently artisan pottery industry. Although overall diversity in specialist-produced pottery assemblages is comparatively low, it does vary within that narrow range as the number and size of production units change. For instance, if the number of workshops increases, it seems likely that the diversity of pottery would increase as well, simply because more potters using different manufacturing techniques and/or decorative styles are involved in the industry. If the scale of some workshops expands, resulting in fewer production units, it seems reasonable that ceramic diversity would decrease as well, because fewer potters or workshops are making pottery. Although no ethnographic studies have been directed specifically toward the question of ceramic diversity in an artisan-dominated industry, two studies provide some helpful clues. In an investigation of traditional pottery-making in Pakistan, Rye and Evans (1976) found that large-scale pottery workshops tended to specialize in a particular vessel type, often producing only one or two kinds, while small workshops tended to make a wider range of vessels for sale to local consumers. A similar observation was made by Vossen (1984, 1986) in a study of pottery-making centers in Morocco and Spain. He found that large-scale operations specialized in specific vessel types, e.g., water storage jars or unglazed cooking pots, which were widely distributed throughout a region. Thus changes in the diversity of locally produced ceramic goods would tend to reflect a shift in the organization of the industry--either toward a more competitive industry or one monopolized by few large-scale producers.

## Diversity in al-Basra's Pottery Assemblages

Although the al-Basra pottery collection appears relatively homogeneous due to the fact that it is almost entirely wheelmade and undecorated, it consists of a variety of distinct wares and forms. As Chapter 5 indicated, al-Basra potters made a wide range of forms from each of the buff-firing clays, i.e., sandy, fine, and very fine, and the cream-firing clays. These locally made wares make up nearly 80% of the collection. Glazed wares are believed to have been imported into the site and constitute only a small proportion of the collection. Although handmade wares were produced locally, they were very likely made by domestic potters for their own use rather than by skilled artisans.

### Functional and Technological Diversity

In this study, I measure ceramic diversity at several hierarchical levels in order to fully understand the changes that took place in the organization of ceramic production at al-Basra during its three occupation phases. The total diversity in each assemblage reflects the total number of locally produced vessel types and the distribution of sherds among those types. While total diversity provides a rough indicator of the overall complexity of an assemblage, it actually consists of two components--functional and technological diversity. Functional diversity represents the variety that exists among functional classes, such as small storage jars, pitchers, serving jars, etc. These classes are distinguished by their general size and shape, features that are related to vessel function. Technological diversity represents the variety among ware-form types, such as cream-colored jars with band rims, cream-colored pitchers, and cream-colored lamps, which are found within each functional group. These technological types are differentiated by their paste composition, rim and neck morphology, and decorative treatment, attributes that are related to manufacturing techniques. In order to isolate the diversity that is relevant to the organization of craft production, it is necessary to look at these aspects separately.

In a very real sense, the diversity in each of these components is caused by different cultural or behavioral factors. The variety of functional classes in an assemblage is due primarily to socio-economic, ethnic, and/or environmental factors. For example, the diversity of functional classes may change because of the arrival of new ethnic groups whose traditional food preparation and serving customs require new kinds of ceramic utensils; it may change because of a growth in social or occupational groups whose tastes, statuses, or jobs require new kinds of vessels or more of one kind, such as elite ceremonial plates or military cooking pots; it may also change because of shifting climatic or ecological conditions which stimulate a need for new kinds of vessels or for more vessels of certain functional classes, such as water storage jars during times of drought.

On the other hand, technological diversity is believed to be primarily the result of production processes. In a general way, technological diversity within functional groups reflects the number of individual potters or workshops producing ceramic goods and the relative proportion of their share of the market, i.e., the amount of competition present among producers during a particular time. As ethnographic studies have shown, traditional potters tend to use raw materials, clay preparation techniques, and forming and firing procedures which differ in some way from that of another workshop or potter. These technological variations are likely to appear to some degree in the finished product. Furthermore, traditional potters or pottery workshops tend to manufacture a similar range of goods, that is, each workshop makes one kind of serving jar, one kind of pitcher, etc., although they may specialize in only one or two functional categories (Balfet 1965; Rye and Evans 1976; Vossen 1984, 1986). The important thing is that workshops seldom produce more than one variety of each functional class. Therefore, by measuring the variability of technological types within each functional group, it is possible to determine in a rough way the relative number of pottery workshops that were involved in the production of a particular functional class and their relative share of the market. The diversity of technological types may change, for example, as the number of small production units increases or as they begin to share the market more equitably. The diversity may decrease as the number of producers declines or as one manufacturing unit expands to monopolize the production of certain classes of pottery.

### Hierarchical Diversity Measure

In order to examine both functional and technological components of ceramic diversity at al-Basra, the study employs a hierarchical diversity measure that was developed in mathematical ecology (Pielou 1977). This measure takes into account the hierarchical nature of certain classifications, such as that of pottery, in which the total diversity of an assemblage is a composite of functional and technological diversity. For example, in a comparison of two pottery assemblages that contain the same number of technological types in the same relative proportions, the diversities of the two assemblages are equal. If, however, in one assemblage all of the types belong to one functional group and in the other assemblage each type belongs to a different functional group, it is reasonable to regard the latter assemblage as the more diverse of the two. This underscores the value of splitting the <u>total</u> diversity of an assemblage into its component parts.

On the basis of a formula presented by Pielou (1977:303-304), <u>total</u> diversity is equal to the sum of functional diversity and weighted average technological diversity within functional groups:

$$H'(f,t) = H'(f) + H'(t)$$

where $H'(f,t)$ is the total diversity; $H'(f)$ is the diversity of functional groups; and $H'(t)$ is the weighted mean diversity of technological types within functional groups. The diversity of functional groups is calculated as

$$H'(f) = -\Sigma p_i \log p_i$$

where $p_i$ is the proportion of the <u>ith</u> functional class within the total assemblage (this formula is the standard Shannon index of diversity). The weighted mean diversity of technological types within functional groups is calculated as

$$H'(t) = -\Sigma p_i H'_i$$

where $H'_i$ is the diversity of technological types <u>within</u> each functional class. The diversity within each functional class is computed as

$$H'_i = -\Sigma p_{ij} \log p_{ij}$$

where $p_{ij}$ is the proportion of technological type $\underline{j}$ within functional class $\underline{i}$.

In these formulas, H' takes on a value of zero when diversity is at its theoretical minimum--i.e., when all the ceramics in an assemblage belong to a single type (or class); as the number of types and the evenness of the distribution of items among types increase, so too does the value of the index. The Shannon index, in particular, has been used in a number of recent archeological studies of artifact diversity (e.g., Conkey 1980; Rice 1981).

In order to adjust for differences in sample size and to assess the significance of changes in diversity from one pottery assemblage to another, I used a computer simulation program developed by Kintigh to determine diversity values (1984a, 1984b). The program calculates the expected diversity for assemblages of different sizes on the basis of an underlying frequency distribution, which in the al-Basra case was the actual distribution of types within each assemblage. Also in the al-Basra case, diversity was calculated on the basis of a mean sample size for all three assemblages, in order to make them comparable. The confidence interval was set at 90%.

## Pottery Used in the Diversity Study

The material chosen for the diversity study consisted of pottery that was most likely produced in the artisan workshops at al-Basra. This material included 4,093 of 5,388 rim sherds, or about 76% of the collection. These sherds belonged to 24 ware-form types from all three urban occupation phases at the site (see Table 8.1 and Figure 8.1). The study excluded all glazed wares, which are believed to have been produced elsewhere in North Africa or Spain, and handmade wares, which were very likely made by domestic potters rather than by artisans. In addition, it excluded types that occur very rarely, i.e., less than 20 examples in the collection, and material from mixed or residual deposits or from those that could not be assigned to a specific phase.

The sherds were grouped into both functional categories and technological categories by occupation phase. Of these sherds, 1,003 belong to Phase 1 assemblages, which represent the earliest occupation at al-Basra; 1,931 sherds belong to Phase 2 assemblages, which are associated with the major rebuilding and second occupation of the site; and 889 sherds come from Phase 3 assemblages, which represent the third and final occupatin at al-Basra. Some 270 sherds from Phase 4 assemblages were not included in the diversity study because they are associated with the post-urban period at the site.

## Functional Diversity: H'(f)

The analysis of functional diversity is based on the classification of the al-Basra pottery into six functional classes, according to vessel size and shape (Table 8.2). These classes included small storage jars, pitchers, lamps, carinated jars, water storage jars, and large storage jars.

As Figure 8.2 shows, functional diversity changes rather dramatically through time. Although the number of functional classes stays the same, the evenness dimension increases from Phase 1 to Phase 2 and levels out in Phase 3. The increase is significant at a 90% confidence level. On the whole, there is a more even distribution of sherds among functional classes in Phases 2 and 3 than there are in Phase 1.

The changes in diversity are clearly reflected in Figure 8.3, which shows the relative frequencies of functional classes by occupation phase. The relative abundance of small storage jars drops sharply from Phase 1 to Phase 3 while the relative abundances of pitchers, carinated jars and, to a lesser extent, lamps increase steadily during

TABLE 8.1

COUNTS AND PERCENTAGES OF MAJOR FORMS, BY WARE GROUP

| Ware Group | Total (No.) | Form 40 (No.) | Form 40 (%) | Form 41 (No.) | Form 41 (%) | Form 42 (No.) | Form 42 (%) | Form 43 (No.) | Form 43 (%) | Form 44 (No.) | Form 44 (%) | Form 48 (No.) | Form 48 (%) | Form 74 (No.) | Form 74 (%) | Form 93 (No.) | Form 93 (%) | Other (No.) | Other (%) |
|---|---|---|---|---|---|---|---|---|---|---|---|---|---|---|---|---|---|---|---|
| Sandy buff | 568 | 126 | 22.2 | 138 | 24.3 | 66 | 11.6 | 138 | 24.3 | 13 | 2.3 | 32 | 5.6 | 24 | 4.2 | 6 | 1.1 | 25 | 4.4 |
| Fine buff | 2,762 | 597 | 21.6 | 555 | 20.1 | 56 | 2.0 | 551 | 19.9 | 142 | 5.1 | 489 | 17.7 | 54 | 2.0 | 176 | 6.4 | 142 | 5.1 |
| Very fine | 779 | 132 | 16.9 | 159 | 20.4 | 4 | 1.0 | 83 | 10.7 | 50 | 6.4 | 226 | 29.0 | 15 | 1.9 | 67 | 8.6 | 43 | 5.5 |
| Cream-color | 648 | 15 | 2.3 | 56 | 8.6 | 9 | 1.4 | 36 | 5.6 | 20 | 3.1 | 297 | 45.8 | 1 | 0.0 | 145 | 22.4 | 69 | 10.7 |
| Coarse | 95 | 2 | 2.1 | 13 | 13.7 | 1 | 1.1 | 7 | 7.4 | 1 | 1.1 | 19 | 20.0 | 4 | 4.2 | 1 | 1.1 | 47 | 50.0 |
| Glazed | 71 | 0 | 0.0 | 3 | 4.2 | 0 | 0.0 | 0 | 0.0 | 0 | 0.0 | 13 | 18.3 | 0 | 0.0 | 21 | 30.0 | 34 | 47.9 |
| Handmade | 465 | 0 | 0.0 | 1 | 0.2 | 1 | 0.2 | 0 | 0.0 | 0 | 0.0 | 0 | 0.0 | 8 | 1.7 | 1 | 0.2 | 454 | 97.6 |
| Total | 5,388 | 872 | | 925 | | 137 | | 815 | | 226 | | 1,076 | | 106 | | 417 | | 814 | |

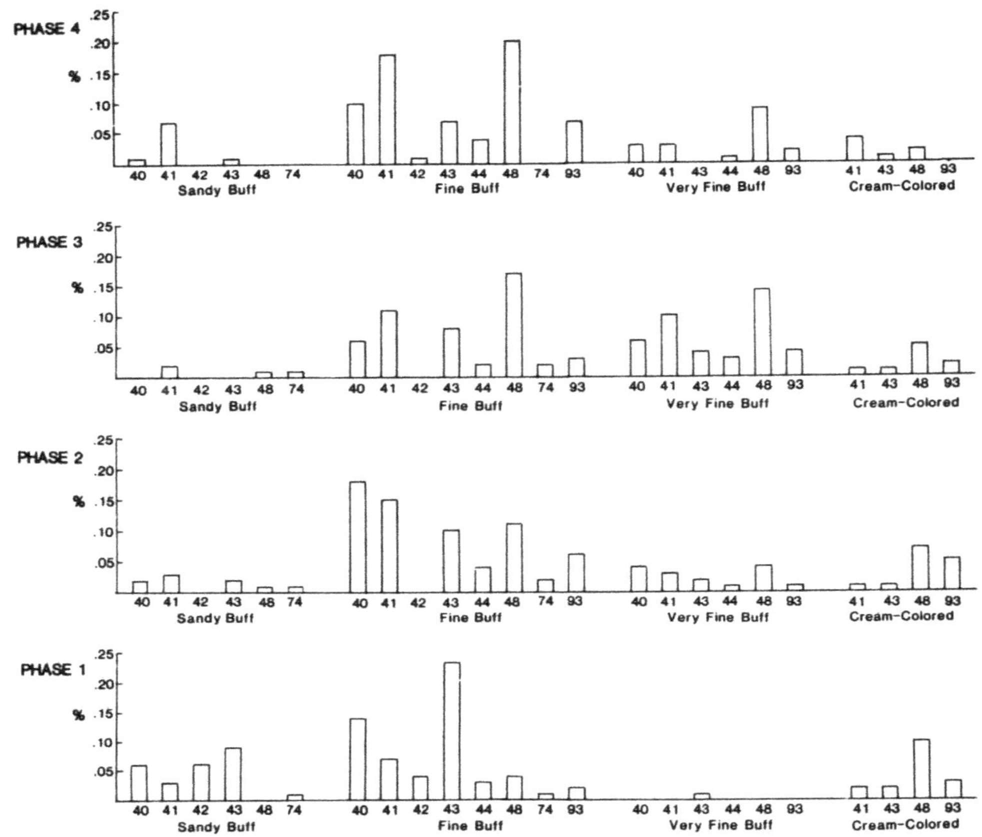


Figure 8.1. Relative frequency of ware-form categories (technological types), by occupation phase.

TABLE 8.2

COUNTS AND PERCENTAGES OF RIM SHERDS IN SIX FUNCTIONAL CLASSES, BY OCCUPATION PHASE

| Occupation Phase | Total Rims (No.) | Small Storage Jars (No.) | Small Storage Jars (%) | Pitchers (No.) | Pitchers (%) | Lamps (No.) | Lamps (%) | Carinated Jars (No.) | Carinated Jars (%) | Large Storage Jars (No.) | Large Storage Jars (%) | Water Storage Jars (No.) | Water Storage Jars (%) |
|---|---|---|---|---|---|---|---|---|---|---|---|---|---|
| 4 Post-urban | 270 | 113 | .42 | 85 | .32 | 24 | .09 | 35 | .13 | 1 | .00 | 12 | .04 |
| 3 Late | 889 | 309 | .35 | 324 | .37 | 83 | .09 | 108 | .12 | 20 | .02 | 45 | .05 |
| 2 Middle | 1,931 | 717 | .37 | 418 | .22 | 213 | .11 | 451 | .23 | 39 | .02 | 93 | .05 |
| 1 Early | 1,003 | 560 | .56 | 137 | .14 | 46 | .05 | 216 | .22 | 16 | .02 | 28 | .03 |
| Total[1] | 4,093 | 1,699 | | 964 | | 366 | | 810 | | 76 | | 178 | |

[1] Total excludes coarse, glazed, and handmade rim sherds and sherds from provenience units that were disturbed or could not be assigned to an occupation phase.

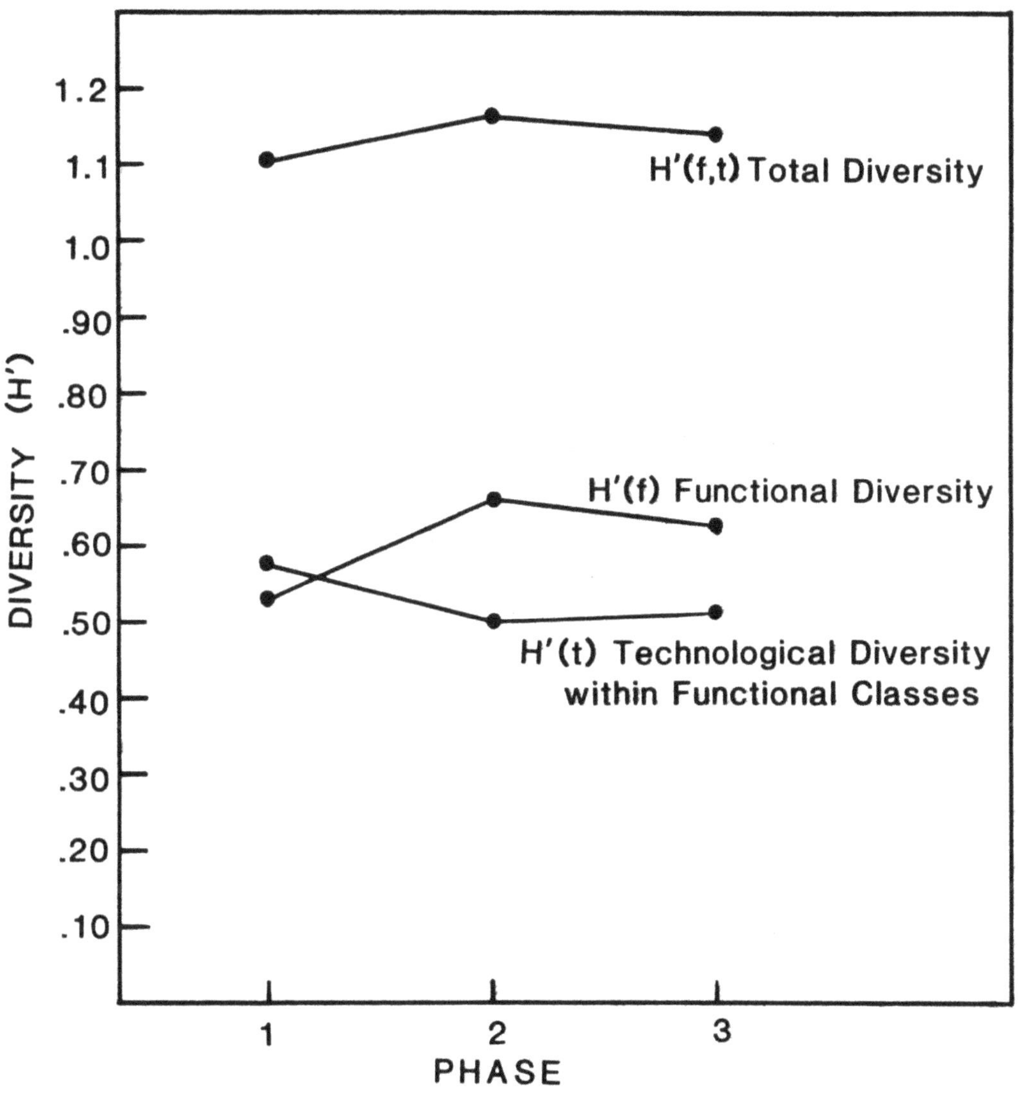


Figure 8.2. Total, functional, and technological diversity, by occupation phase.

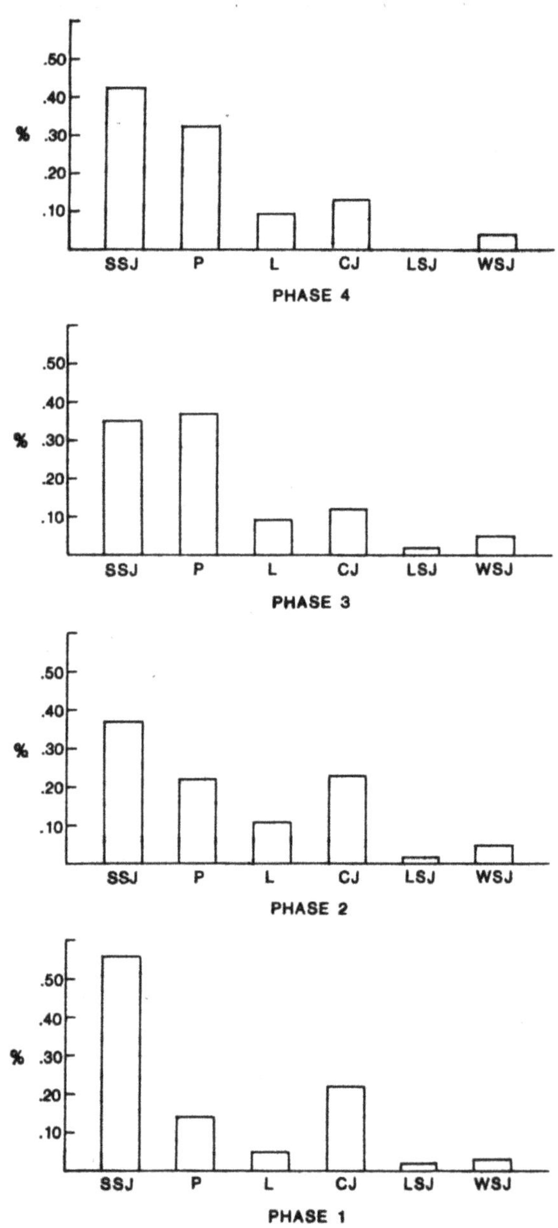


Figure 8.3. Relative frequency of six functional classes, by occupation phase.

the same period. As the charts show, one class--small storage jars--dominates the early phase while three classes--small storage jars, pitchers, and carinated bowls--are more evenly distributed in Phases 2 and 3, resulting in a higher diversity value.

In summary, we can say that, although al-Basra's pottery workshops produced the same range of functional classes for about 300 years, they manufactured varying proportions of these classes. For some yet unknown reasons, local demand for several of these classes, i.e., pitchers, carinated bowls, and lamps, rose during the middle and final occupation phases at the urban center. At the same time, the need for small storage jars declined.

## Technological Diversity: H'(t)

The study of technological diversity is based on the classification of the al-Basra pottery into 24 technological types (ware-form combinations), according to paste composition and rim and neck morphology. They include, for example, sandy buff-firing pitchers, fine buff-firing pitchers, etc. These types were then assigned to one of the six functional classes (Tables 8.3-8.8).

As Figure 8.2 indicates, technological diversity, like functional diversity, changes, but in the opposite direction. The weighted mean diversity of technological types within all functional classes considered together drops from Phase 1 to Phase 2 and then stays the same in Phase 3. Technological diversity on the whole is at its lowest point during the last two occupation phases at al-Basra.

We can get a better idea of the reasons for the overall decline in technological diversity by examining the variety within each functional group considered separately. Within the largest functional class--small storage jars--technological diversity drops in a significant way from Phase 1 to Phase 2 and then evens out in Phase 3 (Figure 8.4a). The decline in diversity occurs at the same time that the overall proportion of small storage jars falls (Figure 8.4b). In terms of production factors, it seems that fewer and/or larger workshops produce small storage jars at a time when consumer demand for this kind of jar is on the wane. The amount of competition among producers who make this kind of jar seems to drop as the demand for it falls.

A different pattern emerges in two other functional groups--pitchers and lamps. As Figure 8.5a illustrates, the diversity of pitchers increases--significantly--from Phase 1 to Phase 2 and then levels out in Phase 3. In a similar fashion, as Figure 8.6a shows, the diversity of lamps increases sharply from Phase 1 to Phase 2 and continues to rise in Phase 3. The general rise in technological diversity within these two vessel classes takes place at the same time that the relative abundance of each class rises (Figures 8.5b and 8.6b). It seems that more and/or smaller workshops begin to produce pitchers and lamps as consumption for both items grows. Competition among potters who make pitchers and lamps increases as the market for these items grows.

TABLE 8.3

COUNTS AND PERCENTAGES OF TECHNOLOGICAL TYPES OF SMALL STORAGE JARS, BY OCCUPATION PHASE

| Occupation Phase | Total (No.) | Small Storage Jars | | | | | | | | | | | | | | | | | | | |
|---|---|---|---|---|---|---|---|---|---|---|---|---|---|---|---|---|---|---|---|---|---|
| | | Sandy 41 (No.) | (%) | Sandy 42 (No.) | (%) | Sandy 43 (No.) | (%) | Fine 41 (No.) | (%) | Fine 42 (No.) | (%) | Fine 43 (No.) | (%) | Very Fine 41 (No.) | (%) | Very Fine 43 (No.) | (%) | Cream 41 (No.) | (%) | Cream 43 (No.) | (%) |
| 4 Post-urban | 113 | 18 | .16 | 0 | .00 | 2 | .02 | 49 | .43 | 2 | .02 | 19 | .17 | 9 | .08 | 1 | .01 | 11 | .10 | 2 | .02 |
| 3 Late | 309 | 14 | .05 | 0 | .00 | 3 | .01 | 93 | .30 | 0 | .00 | 68 | .22 | 85 | .28 | 37 | .12 | 5 | .02 | 4 | .01 |
| 2 Middle | 717 | 56 | .08 | 3 | .00 | 38 | .05 | 298 | .42 | 5 | .01 | 192 | .27 | 59 | .08 | 33 | .05 | 20 | .03 | 13 | .02 |
| 1 Early | 560 | 31 | .06 | 60 | .11 | 85 | .15 | 65 | .12 | 40 | .07 | 233 | .42 | 3 | .01 | 7 | .01 | 19 | .03 | 17 | .03 |
| Total | 1,699 | 119 | | 63 | | 128 | | 505 | | 47 | | 512 | | 156 | | 78 | | 55 | | 36 | |

TABLE 8.4

COUNTS AND PERCENTAGES OF TECHNOLOGICAL TYPES
OF PITCHERS, BY OCCUPATION PHASE

| Occupation Phase | Total (No.) | Pitchers: Sandy 48 (No.) | Sandy 48 (%) | Fine 48 (No.) | Fine 48 (%) | Very Fine 48 (No.) | Very Fine 48 (%) | Cream 48 (No.) | Cream 48 (%) |
|---|---|---|---|---|---|---|---|---|---|
| 4 Post-urban | 85 | 1 | .01 | 54 | .64 | 25 | .29 | 5 | .06 |
| 3 Late | 324 | 11 | .03 | 150 | .46 | 122 | .38 | 41 | .13 |
| 2 Middle | 418 | 11 | .03 | 208 | .50 | 69 | .17 | 130 | .30 |
| 1 Early | 137 | 4 | .03 | 35 | .26 | 0 | .00 | 98 | .72 |
| Total | 964 | 27 | | 447 | | 216 | | 274 | |

TABLE 8.5

COUNTS AND PERCENTAGES OF TECHNOLOGICAL TYPES
OF LAMPS, BY OCCUPATION PHASE

| Occupation Phase | Total (No.) | Lamps: Fine 93 (No.) | Fine 93 (%) | Very Fine 93 (No.) | Very Fine 93 (%) | Cream 93 (No.) | Cream 93 (%) |
|---|---|---|---|---|---|---|---|
| 4 Post-urban | 24 | 19 | .79 | 4 | .17 | 1 | .04 |
| 3 Late | 83 | 27 | .33 | 36 | .43 | 20 | .24 |
| 2 Middle | 213 | 107 | .50 | 19 | .09 | 87 | .41 |
| 1 Early | 46 | 16 | .35 | 1 | .02 | 29 | .63 |
| Total | 366 | 169 | | 60 | | 137 | |

TABLE 8.6

COUNTS AND PERCENTAGES OF TECHNOLOGICAL TYPES OF CARINATED JARS, BY OCCUPATION PHASE

| Occupation Phase | Total (No.) | Carinated Jars: Sandy 40 (No.) | Sandy 40 (%) | Very Fine 40 (No.) | Very Fine 40 (%) | Fine 40 (No.) | Fine 40 (%) |
|---|---|---|---|---|---|---|---|
| 4 Post-urban | 35 | 2 | .06 | 26 | .74 | 7 | .20 |
| 3 Late | 108 | 3 | .03 | 54 | .50 | 51 | .47 |
| 2 Middle | 451 | 38 | .08 | 346 | .77 | 67 | .15 |
| 1 Early | 216 | 74 | .34 | 139 | .64 | 3 | .01 |
| Total | 810 | 117 | | 565 | | 128 | |

TABLE 8.7

COUNTS AND PERCENTAGES OF TECHNOLOGICAL TYPES OF WATER STORAGE JARS, BY OCCUPATION PHASE

| Occupation Phase | Total (No.) | Water Storage Jars: Fine 44 (No.) | Fine 44 (%) | Very Fine 44 (No.) | Very Fine 44 (%) |
|---|---|---|---|---|---|
| 4 Post Urban | 12 | 10 | .83 | 2 | .17 |
| 3 Late | 45 | 19 | .42 | 26 | .58 |
| 2 Middle | 93 | 72 | .77 | 21 | .23 |
| 1 Early | 28 | 27 | .96 | 1 | .04 |
| Total | 178 | 128 | | 50 | |

TABLE 8.8

COUNTS AND PERCENTAGES OF TECHNOLOGICAL TYPES OF LARGE STORAGE JARS, BY OCCUPATION PHASE

| Occupation Phase | Total (No.) | Large Storage Jars: Sandy 74 (No.) | (%) | Fine 74 (No.) | (%) |
|---|---|---|---|---|---|
| 4 Post-urban | 1 | 0 | .00 | 1 | .00 |
| 3 Late | 20 | 4 | .20 | 16 | .80 |
| 2 Middle | 39 | 9 | .23 | 30 | .77 |
| 1 Early | 16 | 10 | .63 | 6 | .38 |
| Total | 76 | 23 | | 53 | |

A third trend appears in another of the major functional groups--carinated jars. As Figure 8.7a illustrates, the diversity of technological types within this functional group changes rather erratically. It falls a little from Phase 1 to Phase 2 and then increases in Phase 3. Overall, technological diversity is at its highest point in the final occupation phase. As Figure 8.7b indicates, however, the relative proportion of carinated jars drops in Phase 3, when it apparently was in low demand. Oddly enough, the variety of types of carinated jars increases at the same time that consumption of this kind of vessel was low. The pattern suggests either that more potters/workshops began to produce carinated jars or that its production became more evenly distributed among workshops during the last occupation period, while the number of vessels that were produced apparently decreased.

The diversity patterns of the remaining functional classes--water storage jars and large storage jars--behave somewhat erratically, primarily because of their very low relative abundances. As Figure 8.8 shows, the diversity of water storage jars increases from Phase 1 to Phase 2 and continues to rise in Phase 3. The diversity is generally highest in the last two occupation phases. In spite of the increase in diversity, the relative proportion of this functional group barely changes. As Figure 8.9 indicates, the diversity of large storage jars declines from Phase 1 to Phases 2 and 3, although not significantly, while the relative proportion of this class remains at the same level throughout all three occupation phases. These trends suggest that the number of potters/workshops making these vessels increased (or decreased) without a concomitant change in the total number of vessels being produced.

In summary, two points about technological diversity need to be emphasized. The first is that the decline in weighted mean technological diversity for all vessel classes considered together largely

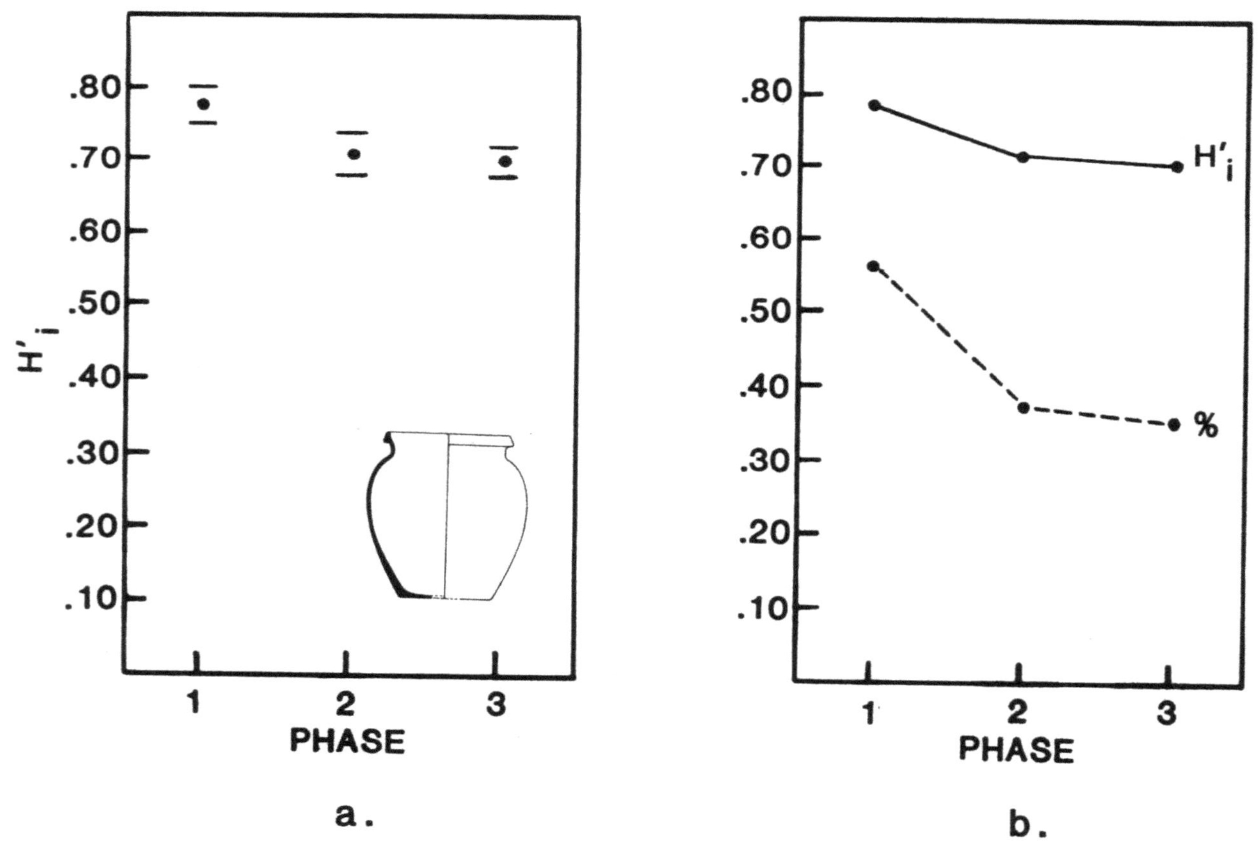


Figure 8.4. Technological diversity and relative frequency of small storage jars, by occupation phase. Diversity values are within a 90% confidence interval.

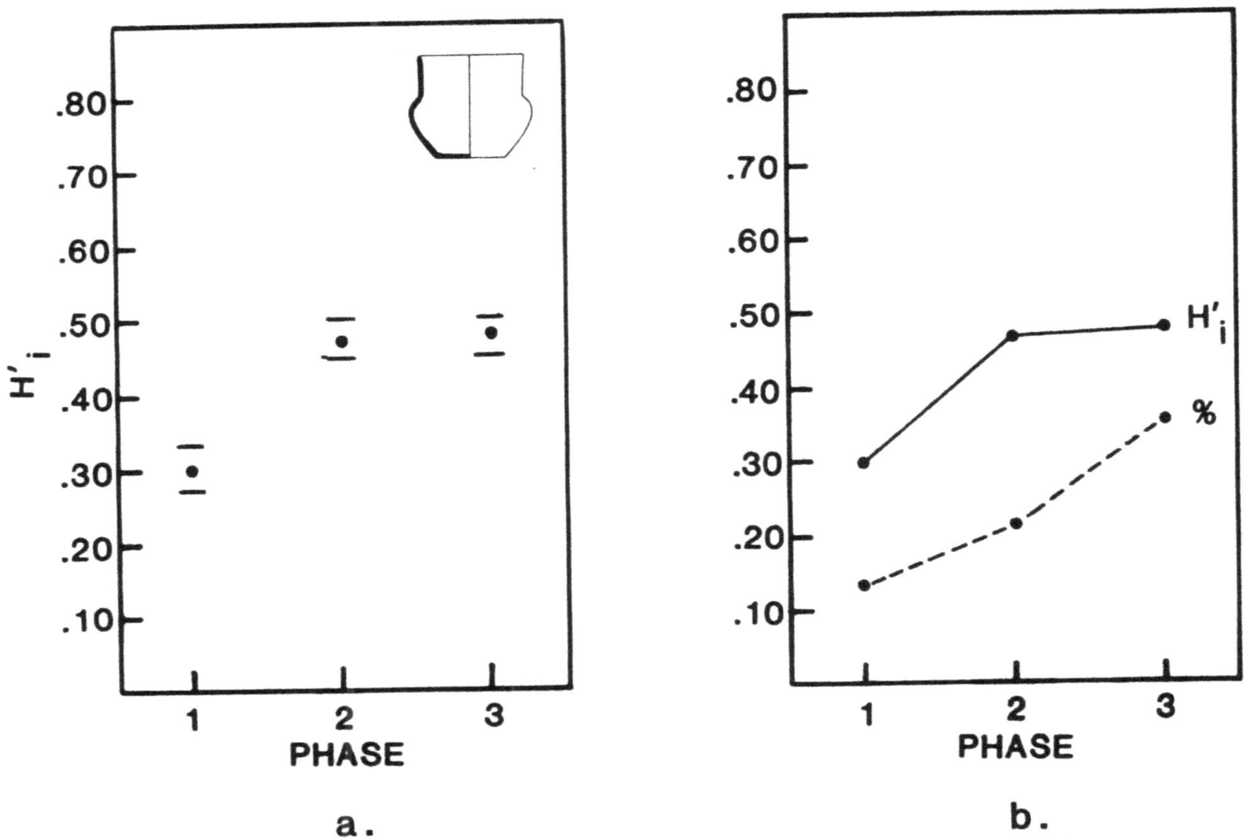


Figure 8.5. Technological diversity and relative frequency of pitchers, by occupation phase. Diversity values are within a 90% confidence interval.

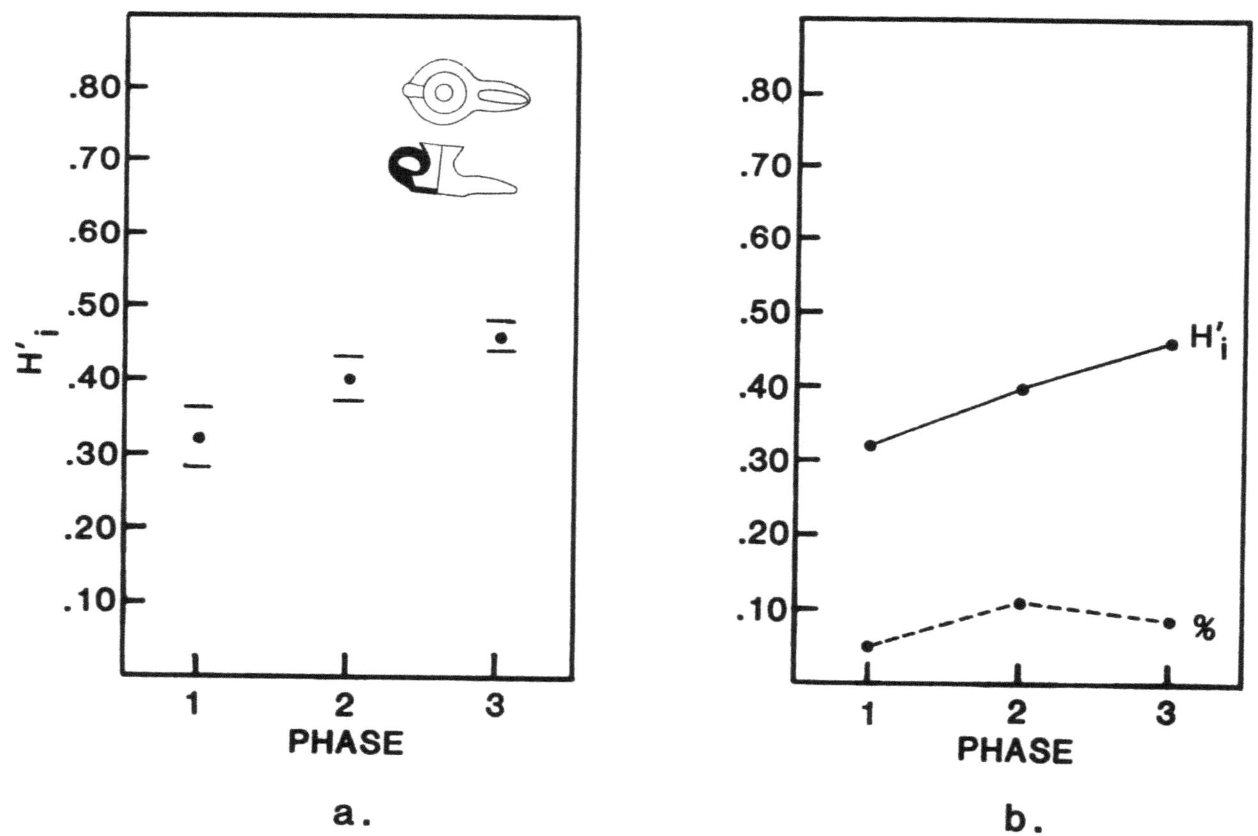


Figure 8.6. Technological diversity and relative frequency of nosed lamps, by occupation phase. Diversity values are within a 90% confidence interval.

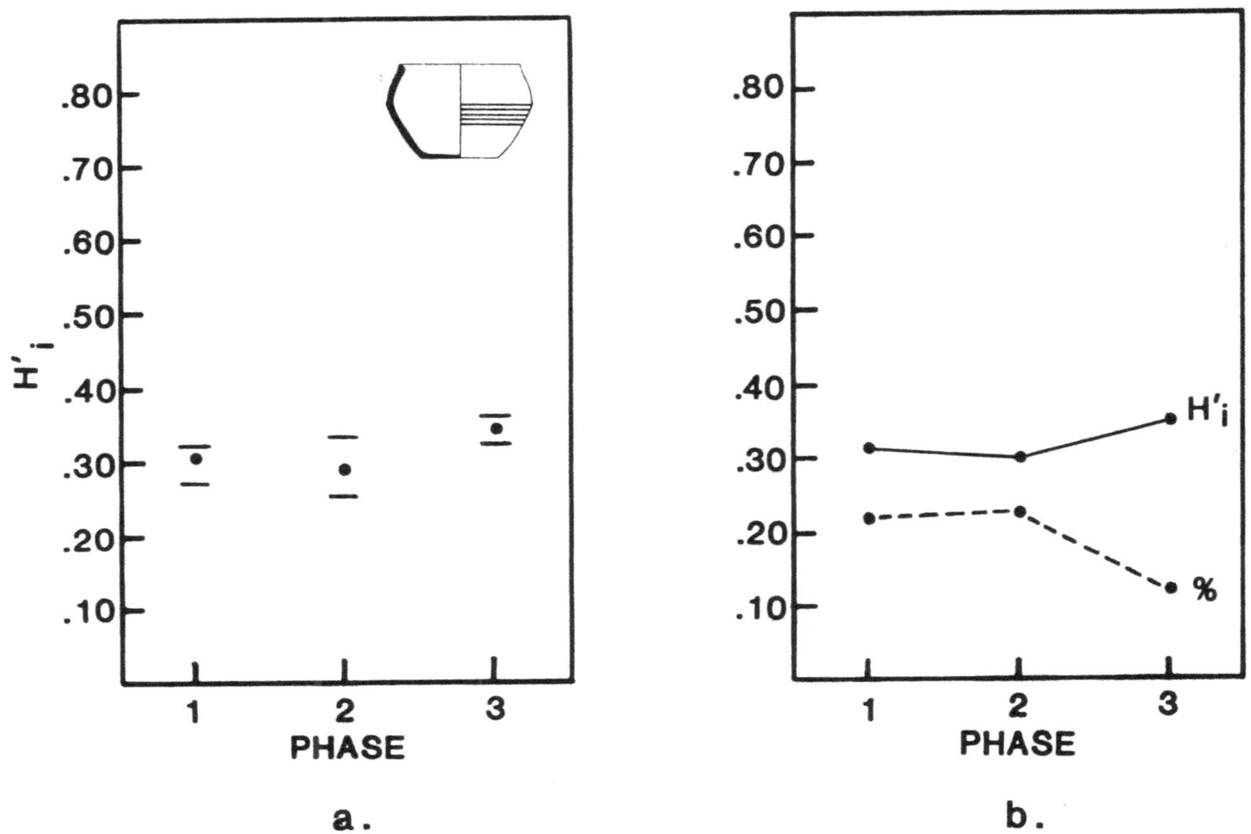


Figure 8.7. Technological diversity and relative frequency of carinated jars, by occupation phase. Diversity values are within a 90% confidence interval.

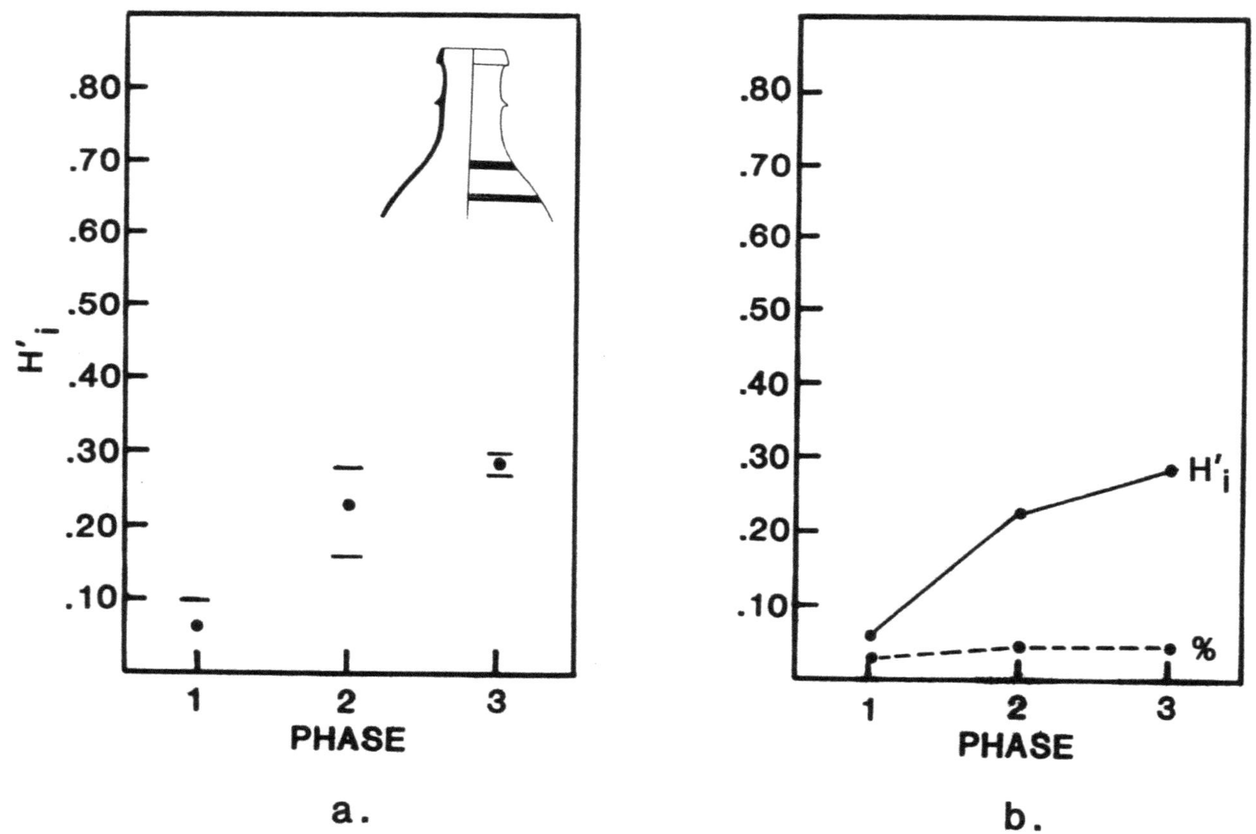


Figure 8.8. Technological diversity and relative frequency of water storage jars, by occupation phase. Diversity values are within a 90% confidence interval.

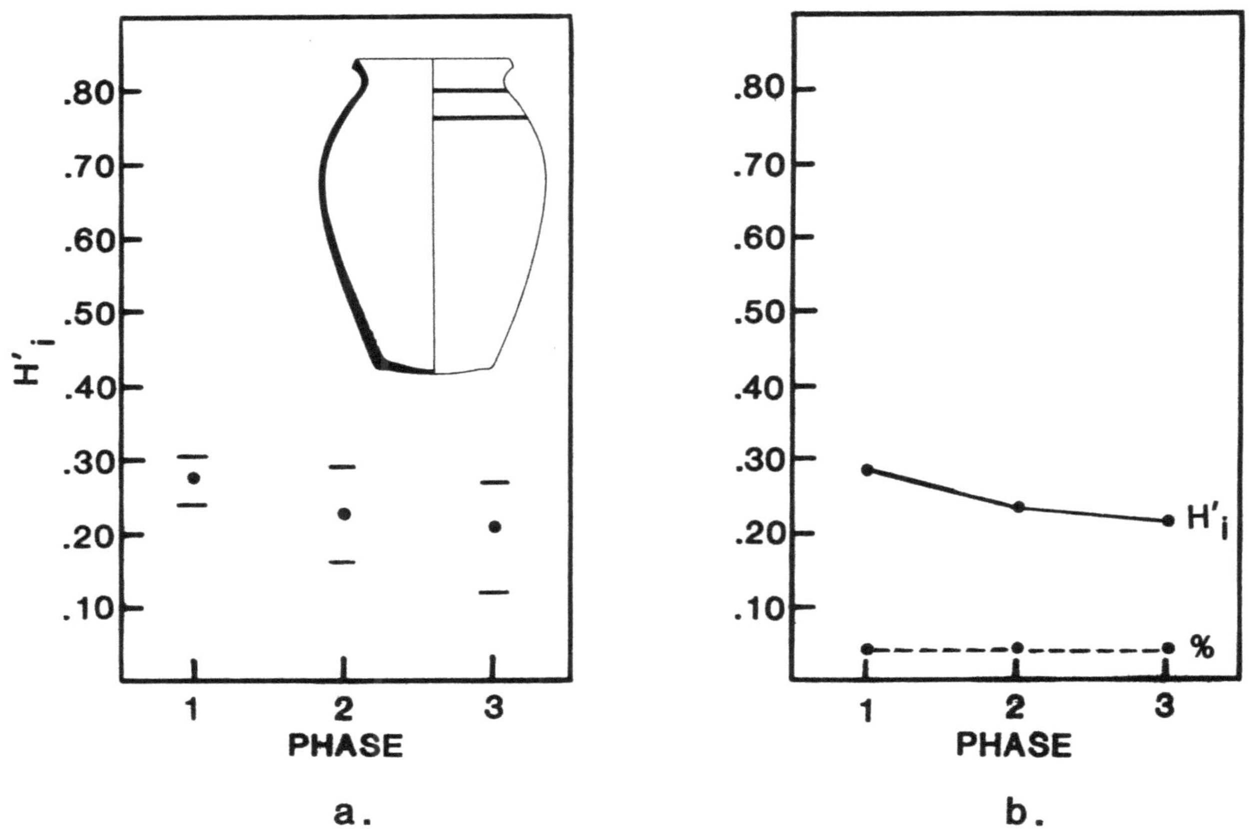


Figure 8.9. Technological diversity and relative frequency of large storage jars, by occupation phase. Diversity values are within a 90% confidence interval.

reflects the drop in the variety of small storage jars, which constitute more than 40% of the al-Basra pottery. A change in the diversity of this large group has a strong impact on the amount of technological diversity as a whole. The second point is that the trends found within each functional class considered separately are what we would expect under normal market conditions, with the possible exception of carinated jars. In most cases, the diversity of the ceramic material increased as the demand for such products increased, suggesting that more potters/workshops entered the fray in an expanding market, and vice versa.

Total Diversity: H'(f,t)

As we noted earlier, total ceramic diversity is equal to the sum of functional and weighted mean technological diversity. As Figure 8.2 shows, total diversity appears to change relatively little through 300 years of occupation at the site, although it rises somewhat from Phase 1 to Phase 2. Of the two components of total diversity, functional diversity probably contributes most to this increase in the middle period. The diversity of this component rises sharply between Phase 1 and Phase 2. At the same time, the weighted mean technological diversity falls.

As the study shows, an examination of total diversity alone, as previous studies of ceramic homogeneity/heterogeneity have attempted (e.g., Feinman 1980), may produce misleading results. While total diversity of the al-Basra assemblages increases somewhat in the middle occupation phase, the diversity of technological types--which provides most of the information about the size and number of pottery workshops--actually decreases during the same period of time. Thus, by separating total diversity into its component parts, we are able to isolate and measure the variability due to production factors independently of that due to functional reasons.

## Assessment of Diversity Results in Light of the Proposed Models

The results of the diversity study are particularly interesting in light of the two models described in Chapter 1. The political model contends that changes in the organization of craft production is primarily the result of administrative factors, e.g., growing political control of production. The free-enterprise model, on the other hand, argues that such changes are motivated primarily by economic pressures, e.g., population growth or movement.

As the analysis of technological diversity shows, there is some indication of a change in the organization of ceramic production during the city's 300-year-history. The shift is from a more diverse and, presumably, a more competitive ceramic industry in the early phase to a less diverse and, by implication, a less competitive production context in the middle and late phases. In other words, relatively fewer and/or larger workshops appear to dominate the pottery industry during the middle and late phases at the medieval site.

In part, these results conform to the expectations of the Feinman-Rice model (see Table 8.9). If government involvement in economic activities increased during the late Idrisid-Fatimid period when administrative influence was potentially highest at al-Basra, competition among potters/workshops would have declined as well, leading to a lower technological diversity of ceramic products. The diversity study indicates that this may have happened. On the other hand, if administrative involvement in production activities decreased during the final period at al-Basra, when the Umayyads and Zenatas ruled northern Morocco, competition among producers--and thus technological diversity--would have increased. The study, however, shows very little change in technological diversity between the middle and late occupation phases.

TABLE 8.9

POLITICAL MODEL OF CERAMIC PRODUCTION AT AL-BASRA

| Phase | Political Centralization | Economic Competition | Expected Technological Diversity |
|---|---|---|---|
| 4 Post-urban | - | - | - |
| 3 Late | Fragmented Umayyad-Zenata confederacy | Strong | High |
| 2 Middle | Idrisid-Fatimid administrative center | Weak | Low |
| 1 Early | Fragmented Idrisid state | Strong | High |

Although it is tempting to attribute the changes in technological diversity to shifts in the degree of political/administrative involvement in the ceramic industry at al-Basra, particularly in the middle phase, an alternative possibility needs to be considered. While a shift to larger and/or fewer pottery workshops in the middle and late phases is fairly clear, the reasons for the change are not as straightforward. It is possible that the shift consisted of a change toward larger commercial workshops operated by private entrepreneurs, in order to meet the rising demand for consumer goods in an expanding market (see Table 8.10). This possibility is supported by the analysis of technological variety within functional groups. As we noted earlier, technological diversity increased or decreased in tandem with a rise or fall in the relative popularity of the item. For example, the diversity of small storage jars dropped at the same time as the production of this vessel class fell. The variety of pitchers, lamps, and possibly water storage jars increased at the same time that demand for them apparently increased. These patterns were not a spurious effect of sample size. Instead, they reflect a normal market situation in which a number of producers were attracted to a locale where the demand

TABLE 8.10

ECONOMIC MODEL OF CERAMIC PRODUCTION AT AL-BASRA

| Phase | Urban Population | Production Units | Expected Technological Diversity |
|---|---|---|---|
| 4 Post-urban | - | - | - |
| 3 Late | City destroyed, rebuilt<br>Population declines | Few, small-scale | Low |
| 2 Middle | City rebuilt<br>Population peaks | Few, large-scale/<br>Many, small-scale | Low/<br>High |
| 1 Early | City settled<br>Population expands | Many, small-scale | High |

for certain products was growing, or vice versa. By contrast, the diversity of carinated jars increased at a time when this vessel class seemed to be in lower demand. It is difficult to postulate why this happened, since it seems to run counter to the normal free-market pattern. Various producers may have continued to make this kind of vessel, despite the fact that local demand for it was in decline. It is conceivable that this vessel was distributed to other towns and cities in the region.

The remarkable fact in this analysis is that in no case did technological diversity drop while the relative abundance of the functional group continue to rise, a situation that one might expect if the government became involved in pottery production--especially mass production--as the Feinman-Rice model postulated. If government interference in pottery production increased during Phase 2 when al-Basra became a provincial administrative center for the Idrisids and Fatimids, one would expect to see a decline in competition among pottery producers at the same time that the scale of production of certain kinds of vessels increased. In other words, the diversity within functional classes produced by state-controlled shops should drop as the relative abundance of the vessel group increased. This did not happen. Instead, diversity within functional groups, such as pitchers and lamps, increased as their relative abundances rose in Phases 2 and 3.

## Summary and Conclusions

The results of the diversity study strongly suggest that the al-Basra pottery was produced by independent artisans in small- and/or large-scale commercial workshops. The industry changed somewhat from one phase to another, as the number of producers increased or decreased or as the scale of production in one or two workshops changed, probably in response to shifts in the size of the urban population. There is, however, no evidence for government involvement in production, either in the patterns found in the diversity study or in the type of pottery that was produced.

As the study suggests, the number and size of artisan workshops involved in the production of pottery varied over time, most likely in response to fluctuations in the urban market. The shift was from a more competitive to a less competitive and, thus, less diverse industry between the early and middle occupation phases. During the middle phase, fewer producers and/or larger workshops began to manufacture utilitarian pottery at al-Basra. Some of these potters seem to have shifted from the manufacture of small storage jars to the production of pitchers and lamps, which seem to have become increasingly popular. In addition, some of these producers began to make a higher quality product, in the form of very fine wares, perhaps indicating a general improvement in the standard of living. These changes coincided with the construction of well-made stone-and-mortar structures throughout the city and, possibly, the massive city walls, and seem to reflect a period of relative growth and prosperity at al-Basra.

# CHAPTER 9

# SUMMARY AND CONCLUSIONS

The al-Basra study has brought together both archeological and documentary evidence in an investigation of craft production an early medieval society. The historical information has been used as a framework for examining and interpreting the archeological record from al-Basra in order to reconstruct the pottery industry in this small, provincial town. The historical data have also been used to define the parameters of certain forms of craft production in early state and imperial societies, in particular administratively-directed pottery production. These have been used to develop an alternative model in order to broaden the scope of future archeological inquiries into the organization of craft production and into the factors that may have been responsible for its transformation.

This chapter reviews the major findings of the research and discusses some future directions for archeological studies of non-agricultural production in complex societies.

## Ceramic Production at al-Basra

As both historical and archeological evidence indicate, pottery production in the early medieval town of al-Basra was undertaken by independent artisans in relatively small workshops and was affected primarily by economic and cultural factors, similar to most pottery industries in the Islamic world.

The ceramic material that was produced at the site consisted almost exclusively of utilitarian wares, i.e., vessels used for storage, cooking, and lighting. The utilitarian pottery was made on a wheel and fired in a kiln and, for the most part, remained undecorated. In the Islamic world, this kind of pottery was generally produced for and sold to local urban and rural inhabitants and was not intended for long-distance exchange. Archeological evidence, such as kiln remains and wasters, from other sites in the Maghreb indicate that pottery was produced in virtually all urban centers. Although some of this production may have consisted of glazed and decorated wares for shipment to distant markets, it is reasonable to conclude that most centers also produced a lower quality utilitarian product for local consumption. The concentration of utilitarian pottery production in early Islamic cities differs somewhat from that found in the late Roman world, where major pottery centers specializing in utilitarian wares were often located in provincial areas, such as Gaul and North Africa. The availability of efficient transportation systems, including roads and waterways, made such decentralized commercial production feasible in the Roman period whereas, during the Islamic era, the lack of roadways and large shipping vessels (Unger 1980) made the long distance movement of common, everyday ceramic vessels impractical and uneconomic.

Both the technology and the organization of the pottery industry at al-Basra were highly stable. As archeological evidence indicates, many of the pottery wares and forms made when the Arabs originally occupied the city continued to be produced up to the time that the city was finally abandoned, some three centuries later. These included fine buff-firing storage jars, pitchers, carinated jars and lamps, which constituted the bulk of the pottery found at al-Basra. They represented a manufacturing process that was highly resistant to change. This process continued to dominate the industry even after the introduction of a new manufacturing technology during the second occupation phase when the city was rebuilt. The new method, which involved either the use of a new clay source or a different clay preparation technique, such as levigation, resulted in an aesthetically more pleasing and possibly more durable vessel. By the third and final occupation phase at al-Basra, however, the very fine buff-firing wares had increased in relative abundance, suggesting that they were in greater demand by the local population than other wares. As the study indicates, however, the introduction of a new manufacturing process in an artisan-dominated, urban industry does not mean that it will be adopted rapidly by other local potters, nor that it will be immediately popular with a broad spectrum of local consumers. Instead, the replacement of one product by another is likely to be fairly gradual. This finding underscores the highly conservative nature of specialized artisans who rely on a relatively complicated technological process for their livelihood.

The al-Basra study also emphasizes the conservative nature of the organization of labor in craft production. As a comparative historical study of Roman and Islamic craft production suggests, the same basic forms of workshop organization persisted in the Mediterranean world for more than a millennium. These similarities included not only the actual workshop arrangements but also the involvement of the state or imperial government in certain kinds of craft production. The archeological evidence from al-Basra suggests that the pottery industry consisted of a few small, independent workshops. The number and size of the workshops seem to have changed somewhat through time but the change was relatively modest. During the early occupation phase, the number of artisans was relatively high and their production scale was fairly modest. During the second and third occupation phases, their number decreased and the scale of production in a few workshops increased, reflecting a decline in competition among the city's potters. Although the reasons for such a change are difficult to ascertain, the ceramic material provides some clues. According to the diversity study, an increase in the number of potters/workshops producing a specific kind of vessel occurred in tandem with an increase in the production of and, presumably, consumer demand for that product, and vice versa. In other words, the industry expanded in response to growing consumer demand for certain kinds of vessels and constricted in terms of production when demand fell. It is possible that a drop in consumer demand for certain kinds of vessels, such as small storage jars, jeopardized those workshops that produced them almost exclusively. This may explain why potters who concentrated on the production of small storage jars using sandy, buffing-firing clays may have gone out of business, or may have switched to a new manufacturing technology, when the demand for those vessels dropped in the second and third occupation phases. On the other hand, potters who produced a wide range of vessels, such as those employing fine buff-firing clays, seem to have held onto a large share of the market. In all cases, however, changes in technology and in the number and scale of production units seem to have been stimulated by economic factors rather than by any involvement on the part of state or imperial administrators.

## Ceramic Industries in Early Urban Societies

On a broader scale, the al-Basra study suggests that artisan pottery industries in other early state and imperial societies may have been, at least in part, entrepreneurial. As both the discussions of Roman and Islamic craft production showed, government and/or administrative involvement in the manufacture of craft products was usually limited to items that the government needed for its own operation. These items almost always included textiles for palace furnishings and diplomatic exchanges, armaments for military troops, and coinage. With few exceptions, i.e., military pottery and elite wares, it did not include ceramic goods.

In both Mediterranean civilizations, ceramic production usually operated as a free-enterprise system, in which small and large workshops were owned and operated by independent artisans. Most of the pottery that they produced was sold to consumers either through local markets for utilitarian products or through middlemen for luxury goods. Only a relatively small proportion of the total pottery produced in these societies was made under commission to or direct control of the state or its agents; this material consisted of military wares, particularly in the Roman period, and highly glazed and decorated luxury goods, such as lustre wares, in the Arab world.

The Feinman-Rice model, which stresses political change as a primary motivation for shifts in ceramic production, does not seem applicable to the particular context of pottery manufacture in pre-industrial Islamic society. Instead, demographic, economic, and cultural factors seem to have had a more direct and lasting impact on ceramic production in the early Islamic period. These factors need to be considered in any reconstruction of ancient pottery production.

## Historical Context and Free-Enterprise Production

The availability of both documentary information and well-provenienced archeological material has allowed us to reconstruct the pottery industry at al-Basra in some detail. This kind of close-up study underscores the shortcomings of systemic models, such as those proposed by Feinman, Rice, and others (e.g., Arnold 1975, 1985; Wright and Johnson 1975), which have come under increasingly critical scrutiny in recent years by various archeological researchers (Adams 1974; Adams 1979; Hodder 1982). Many of the limitations cited by these critics are applicable to the recent studies of craft production by Arnold (1985), Feinman (1980, 1985), and Rice (1981). In general, these studies view cultural phenomena, such as craft production as functional, adaptive processes, which change as the larger cultural system adjusts to a changing environment. For example, Feinman attributed an extension of political control over the economy, including craft production, to periodic food shortages in the Oaxaca Valley (Feinman 1980, 1985). In addition, these studies implicitly view human beings as passive, unconscious beings rather than as active, goal-motivated individuals. According to Feinman, changes in the Oaxaca pottery industry were motivated primarily by political/administrative circumstances rather than by an active response by individual potters/workshops to new economic or demographic conditions. The systemic approach, in fact, minimizes the potential importance of individual artisans in making decisions and taking risks independently. Finally, and most importantly, these studies deemphasize the historical context of pottery production, both in terms of the particular historical development of craft production in a region and of specific cultural attitudes toward ceramic material. Thus, researchers in the Oaxaca Valley have assumed that prehistoric inhabitants of the valley placed a similar value on all ceramic material. The present study has clearly shown that the early Islamic inhabitants of North Africa placed different values on various kinds of pottery. They assigned the values according to a culturally shaped perception of the function, durability, aesthetic, and/or intrinsic value of certain kinds of ceramic vessels. These culturally specific and historically derived values and meanings assigned to pottery and other material goods need to be considered in archeological investigations of ancient societies.

## APPENDIX 1

### AVERAGE SHERD SIZE, AND SHERD AND BONE DENSITIES, BY PROVENIENCE UNIT (LEVEL-LOCUS)

| Phase | Level-Locus | Deposit Type | Average Sherd Size (g) | Sherd Density ($m^3$) | Bone Density ($m^3$) |
|---|---|---|---|---|---|
| 0 | A102-0 | - | 11.7 | - | - |
| 0 | C102-0 | MC | 10.5 | .8 | - |
| 3 | D103-1 | MC | 8.4 | 4.4 | 1.1 |
| 0 | E102-2 | UF | 29.4 | 3.7 | .3 |
| 0 | E103-2 | UF | 27.7 | 1.7 | .5 |
| 4 | F101-0 | TS | 12.8 | 2.0 | 1.4 |
| 2 | F103-0 | GF | 13.8 | 11.8 | 7.6 |
| 2 | F104-0 | UF | 11.1 | 8.8 | 3.9 |
| 2 | F105-0 | KC | 8.5 | 6.7 | 3.0 |
| 2 | F106-1 | KC | 9.2 | 5.0 | 2.2 |
| 2 | F107-1 | KC | 8.4 | 1.9 | 1.0 |
| 2 | F103-3 | GF | 9.0 | 16.4 | 16.0 |
| 2 | F104-3 | UF | 11.2 | 10.9 | 7.6 |
| 2 | F105-3 | KC | 9.9 | 2.4 | 1.7 |
| 2 | F108-4 | KC | 8.4 | 5.0 | .8 |
| 2 | F105-5 | KC | 9.1 | 10.6 | 4.2 |
| 2 | F106-5 | KC | 13.3 | 2.5 | 1.6 |
| 4 | F101-6 | TS | 10.1 | 1.7 | .8 |
| 2 | F102-6 | UF | 17.7 | 3.6 | 2.6 |
| 2 | F103-6 | UF | 18.1 | 10.7 | 6.7 |
| 1 | F104-6 | SS | 11.9 | - | - |
| - | F104-7 | - | 16.5 | - | - |
| 2 | F206-1 | GF | 17.1 | 1.8 | 0.0 |
| 4 | F301-0 | TS | 9.1 | 1.1 | .3 |
| 3 | F302-0 | MC | 9.6 | 5.7 | 2.3 |
| 3 | F303-0 | F1 | 14.0 | 11.4 | 5.2 |
| 2 | F304-0 | GF | 24.6 | 2.2 | 1.4 |
| 2 | F305-0 | GF | 19.4 | 2.5 | 1.3 |
| 2 | F306-0 | F2 | 22.8 | 3.2 | 3.1 |
| 1 | F307-0 | KC | 20.7 | 2.8 | 1.9 |
| 1 | F308-0 | KC | 23.9 | 6.2 | 2.1 |
| 1 | F309-0 | KC | 24.2 | 3.4 | 1.6 |
| 4 | F301-1 | TS | 14.2 | 1.1 | .3 |
| 3 | F302-1 | F1 | 11.7 | 5.9 | 2.8 |
| 2 | F303-1 | GF | 17.4 | 3.7 | 1.7 |
| 4 | F301-2 | TS | 18.4 | 1.7 | .5 |
| 2 | F302-2 | GF | 17.7 | 3.9 | 1.0 |
| 2 | F303-2 | GF | 18.3 | 1.1 | .8 |
| 2 | F304-2 | F2 | 17.6 | 6.3 | 3.3 |
| 4 | F301-3 | TS | 14.0 | .9 | 2.9 |
| 3 | F302-3 | UF | 14.9 | 1.9 | 8.9 |
| 3 | F303-3 | UF | 17.1 | 4.8 | 1.7 |
| 2 | F304-3 | GF | 16.8 | 4.2 | 2.1 |

| Phase | Level-Locus | Deposit Type | Average Sherd Size (g) | Sherd Density ($m^3$) | Bone Density ($m^3$) |
|---|---|---|---|---|---|
| 2 | F305-3 | UF | 19.8 | 6.3 | 1.9 |
| 1 | F306-3 | KC | 24.0 | 4.5 | 1.3 |
| 1 | F307-3 | KC | 19.7 | 4.4 | 0.0 |
| 1 | F308-3 | KC | 24.9 | - | - |
| 1 | F407-0 | UF | 12.2 | - | - |
| 4 | F501-0 | TS | 7.8 | 4.7 | 0.1 |
| 4 | F502-0 | TS | 9.6 | 2.2 | 0.1 |
| 3 | F503-0 | MC | 12.1 | 4.4 | 1.1 |
| 3 | F504-0 | F1 | 9.3 | 6.0 | 2.0 |
| 2 | F505-0 | UF | 9.8 | 15.2 | 10.2 |
| 2 | F506-0 | UF | 11.1 | 19.2 | 10.6 |
| 2 | F507-0 | UF | 12.2 | 27.1 | 14.0 |
| 1 | F508-0 | UF | 13.2 | 6.1 | 4.0 |
| 1 | F509-0 | UF | 12.4 | 6.6 | 5.0 |
| 1 | F510-0 | UF | 11.7 | 12.3 | 5.5 |
| 1 | F511-0 | UF | 11.2 | 8.6 | 2.5 |
| 1 | F512-0 | UF | 11.0 | 12.0 | 3.8 |
| 1 | F513-0 | SS | 9.8 | 3.6 | 1.4 |
| 1 | F514-0 | SS | 10.9 | 2.6 | 0.1 |
| 1 | F515-0 | SS | 17.1 | 0.9 | 0.0 |
| 1 | F516-0 | SS | 5.0 | 0.3 | 0.0 |
| 3 | F505-1 | F1 | 13.1 | 8.4 | 6.0 |
| 0 | F506-1 | CF | 10.4 | 9.5 | 12.5 |
| 2 | F507-1 | UF | 11.7 | 9.2 | 3.5 |
| 3 | F603-0 | MC | 10.6 | 11.7 | 9.7 |
| 2 | F607-0 | UF | 13.7 | 14.5 | 6.3 |
| 4 | F603-1 | F2 | 12.3 | 5.4 | 2.7 |
| 3 | F604-1 | UF | 11.2 | 7.2 | 3.8 |
| 3 | F605-1 | UF | 13.0 | 6.1 | 5.6 |
| 4 | G100-0 | TS | 1.6 | - | - |
| 4 | G101-0 | TS | 6.6 | 1.3 | .3 |
| 4 | G102-1 | TS | 9.4 | 1.5 | 0.7 |
| 3 | G103-1 | MC | 10.2 | 1.1 | 1.6 |
| 3 | G104-1 | MC | 11.5 | 1.0 | .6 |
| 3 | G105-1 | F1 | 10.3 | 1.9 | .4 |
| 2 | G106-1 | UF | 14.7 | 6.6 | 3.3 |
| 2 | G107-1 | UF | 16.8 | 8.7 | 3.5 |
| 2 | G109-1 | UF | 12.5 | 6.8 | 2.1 |
| 0 | G110-1 | CF | 17.1 | 17.0 | 6.9 |
| 1 | G111-1 | UF | 18.8 | 9.2 | 5.2 |
| 1 | G112-1 | UF | 16.6 | 13.8 | 1.8 |
| 1 | G113-1 | UF | 13.1 | 9.1 | 1.1 |
| 1 | G114-1 | SS | 21.8 | 3.0 | .3 |
| 4 | G102-2 | TS | 9.5 | 1.9 | 1.0 |
| 3 | G103-2 | MC | 8.6 | 2.7 | .9 |
| 3 | G104-2 | MC | 12.2 | 1.4 | 1.0 |
| 3 | G105-2 | F1 | 11.3 | 3.8 | 2.3 |
| 3 | G106-2 | UF | 14.8 | 7.0 | 4.4 |
| 2 | G107-2 | UF | 14.7 | 3.2 | 2.4 |
| 2 | G108-2 | UF | 14.8 | 4.1 | 2.0 |

| Phase | Level-Locus | Deposit Type | Average Sherd Size (g) | Sherd Density ($m^3$) | Bone Density ($m^3$) |
|---|---|---|---|---|---|
| 2 | G109-2 | CF | 16.3 | 19.3 | 4.2 |
| 2 | G110-2 | CF | 11.5 | 3.5 | .9 |
| 1 | G111-2 | UF | 17.6 | 8.4 | 3.3 |
| 1 | G112-2 | UF | 21.2 | 10.3 | 1.8 |
| 1 | G113-2 | CF | 20.2 | 24.1 | 4.4 |
| 1 | G114-2 | SS | 18.5 | 7.9 | .6 |
| 1 | G115-2 | SS | 18.3 | 1.4 | .1 |
| 4 | G201-0 | TS | 7.1 | 2.3 | .1 |
| 4 | G202-0 | TS | 6.1 | 5.6 | .5 |
| 3 | G203-1 | MC | 5.4 | 6.4 | 1.2 |
| 3 | G204-1 | MC | 13.9 | 8.2 | .9 |
| 3 | G206-1 | F1 | 12.3 | 15.6 | 1.9 |
| 3 | G203-2 | MC | 8.4 | 6.2 | 2.3 |
| 3 | G204-2 | MC | 23.4 | 3.3 | .6 |
| 3 | G205-2 | F1 | 17.4 | 6.4 | .3 |
| 3 | G206-2 | F1 | 20.5 | 6.5 | .4 |
| 0 | G207-2 | CF | 9.2 | 4.1 | 0.0 |
| 2 | G208-2 | UF | 14.9 | 4.6 | .9 |
| 2 | G209-2 | F2 | 20.1 | 7.0 | 1.0 |
| 2 | G210-2 | UF | 21.1 | 13.8 | 0.0 |
| 1 | G211-2 | CF | 19.1 | 20.3 | 2.1 |
| 1 | G212-2 | CF | 15.0 | 20.0 | 4.6 |
| 1 | G213-2 | F2 | 13.0 | 3.4 | 1.9 |
| 3 | G207-3 | MC | 12.5 | .9 | .3 |
| 2 | G208-3 | F1 | 14.1 | 1.4 | 1.0 |
| 2 | G209-3 | UF | 13.3 | 5.1 | 2.8 |
| 1 | G211-3 | CF | 19.5 | 17.3 | 5.9 |
| 1 | G212-3 | MC | 18.8 | 4.2 | 1.7 |
| 1 | G213-3 | F1 | 13.2 | 6.9 | 1.7 |
| 1 | G214-3 | CF | 16.6 | 3.5 | 1.5 |
| 1 | G215-3 | SS | 8.5 | .9 | 1.3 |
| 1 | G216-3 | SS | 6.6 | .7 | 0.0 |
| 3 | G302-0 | F1 | - | 2.3 | 1.1 |
| 4 | G302-2 | UF | 7.9 | 2.3 | 1.3 |
| 3 | G303-2 | F1 | 6.9 | 4.8 | 3.3 |
| 4 | G301-3 | TS | 7.9 | 6.2 | 2.3 |
| 4 | G302-3 | UF | 9.0 | 5.3 | 2.9 |
| 3 | G303-3 | F1 | 12.7 | 2.1 | 1.2 |
| 4 | H101-0 | TS | 9.4 | 1.6 | .2 |
| 0 | H102-0 | UF | - | 6.7 | 1.3 |
| 0 | H104-0 | UF | 18.8 | 2.1 | 2.0 |
| 0 | I204-2 | UF | 22.4 | 5.2 | .5 |
| 0 | I206-2 | UF | 23.5 | 2.6 | .8 |
| 0 | I207-2 | SS | 16.4 | 1.4 | .9 |

APPENDIX 2

PRESENCE/ABSENCE OF METAL AND GLASS ARTIFACTS, BY PROVENIENCE UNIT (LEVEL-LOCUS)

| Phase | Level-Locus | Deposit Type | Slag | Iron | Nails | Copper | Glass | Lead |
|---|---|---|---|---|---|---|---|---|
| 0 | A102-0 | - | 0 | 0 | 0 | 1 | 0 | 0 |
| 0 | C102-0 | MC | 0 | 1 | 0 | 0 | 0 | 0 |
| 3 | D103-1 | MC | 0 | 1 | 0 | 0 | 0 | 0 |
| 0 | E102-2 | UF | 0 | 0 | 0 | 0 | 0 | 0 |
| 0 | E103-2 | UF | 0 | 0 | 0 | 0 | 0 | 0 |
| 4 | F101-0 | TS | 0 | 0 | 0 | 0 | 0 | 0 |
| 2 | F103-0 | GF | 1 | 1 | 0 | 0 | 0 | 0 |
| 2 | F104-0 | UF | 0 | 1 | 0 | 1 | 0 | 0 |
| 2 | F105-0 | KC | 0 | 1 | 0 | 0 | 0 | 0 |
| 2 | F106-1 | KC | 0 | 1 | 0 | 0 | 0 | 0 |
| 2 | F107-1 | KC | 0 | 0 | 0 | 0 | 1 | 0 |
| 2 | F103-3 | GF | 0 | 1 | 0 | 1 | 0 | 0 |
| 2 | F104-3 | UF | 0 | 0 | 0 | 0 | 0 | 0 |
| 2 | F105-3 | KC | 0 | 0 | 0 | 1 | 0 | 0 |
| 2 | F108-4 | KC | 0 | 0 | 0 | 0 | 1 | 0 |
| 2 | F105-5 | KC | 0 | 1 | 0 | 1 | 0 | 0 |
| 2 | F106-5 | KC | 0 | 1 | 0 | 0 | 0 | 0 |
| 4 | F101-6 | TS | 0 | 0 | 0 | 0 | 0 | 0 |
| 2 | F102-6 | UF | 1 | 1 | 0 | 1 | 0 | 0 |
| 2 | F103-6 | UF | 0 | 0 | 0 | 0 | 0 | 0 |
| 1 | F104-6 | SS | 0 | 0 | 0 | 1 | 0 | 0 |
| - | F104-7 | - | 0 | 0 | 0 | 0 | 0 | 0 |
| 2 | F206-1 | GF | 0 | 1 | 0 | 0 | 0 | 0 |
| 4 | F301-0 | TS | 1 | 0 | 0 | 0 | 0 | 0 |
| 3 | F302-0 | MC | 0 | 1 | 0 | 0 | 0 | 0 |
| 3 | F303-0 | F1 | 0 | 1 | 0 | 0 | 0 | 0 |
| 2 | F304-0 | GF | 0 | 1 | 0 | 0 | 0 | 1 |
| 2 | F305-0 | GF | 0 | 0 | 0 | 0 | 0 | 0 |
| 2 | F306-0 | F2 | 0 | 1 | 0 | 1 | 0 | 0 |
| 1 | F307-0 | KC | 0 | 1 | 0 | 1 | 0 | 0 |
| 1 | F308-0 | KC | 0 | 1 | 0 | 0 | 0 | 0 |
| 1 | F309-0 | KC | 0 | 0 | 0 | 0 | 0 | 0 |
| 4 | F301-1 | TS | 0 | 1 | 0 | 0 | 0 | 0 |
| 3 | F302-1 | F1 | 0 | 1 | 0 | 0 | 0 | 0 |
| 2 | F303-1 | GF | 0 | 0 | 0 | 0 | 0 | 0 |
| 4 | F301-2 | TS | 0 | 1 | 0 | 1 | 0 | 0 |
| 2 | F302-2 | GF | 0 | 1 | 0 | 0 | 0 | 0 |
| 2 | F303-2 | GF | 0 | 0 | 0 | 0 | 0 | 0 |
| 2 | F304-2 | F2 | 0 | 0 | 0 | 0 | 0 | 0 |
| 4 | F301-3 | TS | 0 | 1 | 0 | 0 | 0 | 0 |
| 3 | F302-3 | UF | 0 | 1 | 0 | 0 | 0 | 0 |
| 3 | F303-3 | UF | 0 | 1 | 0 | 0 | 0 | 0 |
| 2 | F304-3 | GF | 0 | 0 | 0 | 0 | 0 | 0 |
| 2 | F305-3 | UF | 0 | 1 | 1 | 0 | 0 | 0 |
| 1 | F306-3 | KC | 1 | 1 | 0 | 0 | 1 | 0 |

| Phase | Level-Locus | Deposit Type | Slag | Iron | Nails | Copper | Glass | Lead |
|---|---|---|---|---|---|---|---|---|
| 1 | F307-3 | KC | 0 | 1 | 0 | 0 | 0 | 0 |
| 1 | F308-3 | KC | 0 | 0 | 0 | 0 | 1 | 0 |
| 1 | F407-0 | UF | 0 | 0 | 0 | 0 | 0 | 0 |
| 4 | F501-0 | TS | 0 | 0 | 0 | 0 | 0 | 0 |
| 4 | F502-0 | TS | 0 | 0 | 0 | 0 | 0 | 0 |
| 3 | F503-0 | MC | 0 | 1 | 0 | 1 | 0 | 0 |
| 3 | F504-0 | F1 | 0 | 0 | 1 | 0 | 0 | 1 |
| 2 | F505-0 | UF | 0 | 1 | 0 | 0 | 0 | 0 |
| 2 | F506-0 | UF | 0 | 1 | 0 | 1 | 1 | 0 |
| 2 | F507-0 | UF | 0 | 1 | 0 | 0 | 0 | 0 |
| 1 | F508-0 | UF | 0 | 1 | 0 | 0 | 0 | 0 |
| 1 | F509-0 | UF | 0 | 1 | 0 | 0 | 0 | 0 |
| 1 | F510-0 | UF | 1 | 1 | 0 | 0 | 0 | 0 |
| 1 | F511-0 | UF | 0 | 1 | 0 | 1 | 1 | 0 |
| 1 | F512-0 | UF | 1 | 0 | 0 | 0 | 0 | 0 |
| 1 | F513-0 | SS | 1 | 0 | 0 | 0 | 1 | 0 |
| 1 | F514-0 | SS | 0 | 0 | 0 | 0 | 0 | 0 |
| 1 | F515-0 | SS | 0 | 0 | 0 | 0 | 0 | 0 |
| 1 | F516-0 | SS | 0 | 0 | 0 | 0 | 0 | 0 |
| 3 | F505-1 | F1 | 0 | 0 | 0 | 1 | 0 | 0 |
| 0 | F506-1 | CF | 0 | 1 | 0 | 1 | 1 | 0 |
| 2 | F507-1 | UF | 0 | 0 | 0 | 1 | 0 | 0 |
| 3 | F603-0 | MC | 1 | 0 | 1 | 0 | 0 | 0 |
| 2 | F607-0 | UF | 0 | 0 | 0 | 0 | 0 | 0 |
| 4 | F603-1 | F2 | 0 | 0 | 1 | 0 | 0 | 0 |
| 3 | F604-1 | UF | 0 | 0 | 0 | 0 | 0 | 0 |
| 3 | F605-1 | UF | 0 | 1 | 1 | 0 | 0 | 0 |
| 4 | G100-0 | TS | 0 | 0 | 0 | 0 | 0 | 0 |
| 4 | G101-0 | TS | 0 | 1 | 0 | 1 | 0 | 0 |
| 4 | G102-1 | TS | 0 | 0 | 0 | 0 | 0 | 0 |
| 3 | G103-1 | MC | 0 | 1 | 0 | 1 | 0 | 0 |
| 3 | G104-1 | MC | 0 | 1 | 0 | 1 | 1 | 0 |
| 3 | G105-1 | F1 | 0 | 1 | 0 | 1 | 1 | 0 |
| 2 | G106-1 | UF | 0 | 1 | 0 | 1 | 1 | 0 |
| 2 | G107-1 | UF | 0 | 1 | 0 | 0 | 1 | 0 |
| 2 | G109-1 | UF | 0 | 0 | 0 | 0 | 0 | 0 |
| 0 | G110-1 | CF | 0 | 1 | 0 | 0 | 0 | 0 |
| 1 | G111-1 | UF | 0 | 0 | 0 | 0 | 0 | 0 |
| 1 | G112-1 | UF | 0 | 0 | 0 | 0 | 0 | 0 |
| 1 | G113-1 | UF | 0 | 0 | 0 | 0 | 0 | 0 |
| 1 | G114-1 | SS | 0 | 0 | 0 | 0 | 0 | 0 |
| 4 | G102-2 | TS | 0 | 0 | 0 | 0 | 0 | 0 |
| 3 | G103-2 | MC | 0 | 1 | 0 | 1 | 0 | 0 |
| 3 | G104-2 | MC | 0 | 0 | 0 | 0 | 0 | 0 |
| 3 | G105-2 | F1 | 0 | 1 | 0 | 1 | 1 | 0 |
| 3 | G106-2 | UF | 0 | 1 | 0 | 1 | 1 | 0 |
| 2 | G107-2 | UF | 0 | 1 | 0 | 0 | 1 | 0 |
| 2 | G108-2 | UF | 0 | 0 | 0 | 0 | 1 | 0 |
| 2 | G109-2 | CF | 0 | 0 | 0 | 0 | 1 | 0 |
| 2 | G110-2 | CF | 0 | 0 | 0 | 0 | 1 | 0 |
| 1 | G111-2 | UF | 0 | 0 | 0 | 0 | 1 | 0 |

| Phase | Level-Locus | Deposit Type | Slag | Iron | Nails | Copper | Glass | Lead |
|---|---|---|---|---|---|---|---|---|
| 1 | G112-2 | UF | 0 | 0 | 0 | 0 | 0 | 0 |
| 1 | G113-2 | CF | 0 | 0 | 0 | 0 | 1 | 0 |
| 1 | G114-2 | SS | 0 | 0 | 0 | 0 | 0 | 0 |
| 1 | G115-2 | SS | 0 | 0 | 0 | 0 | 1 | 0 |
| 4 | G201-0 | TS | 0 | 0 | 0 | 0 | 0 | 0 |
| 4 | G202-0 | TS | 0 | 0 | 0 | 1 | 0 | 0 |
| 3 | G203-1 | MC | 0 | 0 | 0 | 1 | 0 | 0 |
| 3 | G204-1 | MC | 0 | 0 | 0 | 1 | 0 | 0 |
| 3 | G206-1 | F1 | 0 | 0 | 1 | 0 | 0 | 0 |
| 3 | G203-2 | MC | 0 | 0 | 0 | 1 | 1 | 0 |
| 3 | G204-2 | MC | 0 | 0 | 1 | 1 | 1 | 0 |
| 3 | G205-2 | F1 | 0 | 0 | 1 | 1 | 1 | 0 |
| 3 | G206-2 | F1 | 0 | 0 | 1 | 0 | 1 | 0 |
| 0 | G207-2 | CF | 0 | 0 | 0 | 0 | 1 | 0 |
| 2 | G208-2 | UF | 0 | 0 | 0 | 0 | 0 | 0 |
| 2 | G209-2 | F2 | 0 | 0 | 0 | 0 | 0 | 0 |
| 2 | G210-2 | UF | 0 | 0 | 0 | 0 | 1 | 0 |
| 1 | G211-2 | CF | 0 | 0 | 0 | 0 | 0 | 0 |
| 1 | G212-2 | CF | 0 | 0 | 0 | 0 | 1 | 0 |
| 1 | G213-2 | F2 | 0 | 1 | 0 | 0 | 1 | 0 |
| 3 | G207-3 | MC | 0 | 0 | 0 | 0 | 0 | 0 |
| 2 | G208-3 | F1 | 0 | 0 | 0 | 0 | 0 | 0 |
| 2 | G209-3 | UF | 0 | 0 | 1 | 0 | 1 | 0 |
| 1 | G211-3 | CF | 0 | 1 | 0 | 0 | 1 | 0 |
| 1 | G212-3 | MC | 0 | 1 | 0 | 0 | 1 | 0 |
| 1 | G213-3 | F1 | 0 | 0 | 0 | 0 | 0 | 0 |
| 1 | G214-3 | CF | 0 | 0 | 0 | 0 | 0 | 0 |
| 1 | G215-3 | SS | 0 | 0 | 0 | 0 | 1 | 0 |
| 1 | G216-3 | SS | 0 | 0 | 0 | 0 | 0 | 0 |
| 3 | G302-0 | F1 | 0 | 0 | 0 | 0 | 0 | 0 |
| 4 | G302-2 | UF | 0 | 1 | 0 | 0 | 0 | 0 |
| 3 | G303-2 | F1 | 0 | 1 | 0 | 0 | 1 | 0 |
| 4 | G301-3 | TS | 0 | 1 | 0 | 0 | 0 | 0 |
| 4 | G302-3 | UF | 1 | 1 | 0 | 1 | 0 | 0 |
| 3 | G303-3 | F1 | 0 | 1 | 0 | 1 | 0 | 0 |
| 4 | H101-0 | TS | 0 | 0 | 0 | 0 | 0 | 0 |
| 0 | H102-0 | UF | 0 | 0 | 0 | 0 | 0 | 0 |
| 0 | H104-0 | UF | 0 | 1 | 0 | 0 | 0 | 0 |
| 0 | I204-2 | UF | 0 | 0 | 0 | 0 | 0 | 0 |
| 0 | I206-2 | UF | 0 | 0 | 1 | 0 | 0 | 0 |
| 0 | I207-2 | SS | 0 | 0 | 1 | 0 | 0 | 0 |

APPENDIX 3

COUNTS OF AL-BASRA POTTERY WARES,
BY PROVENIENCE UNIT (LEVEL-LOCUS)

| Phase | Level-Locus | Total | Fine | Sandy | Cream | Very Fine | Glaze | Hand-made | Coarse |
|---|---|---|---|---|---|---|---|---|---|
| 0 | A102-0 | 28 | 22 | 1 | 0 | 5 | 0 | 0 | 0 |
| 0 | C102-0 | 7 | 7 | 0 | 0 | 0 | 0 | 0 | 0 |
| 3 | D103-1 | 57 | 9 | 0 | 5 | 37 | 1 | 5 | 0 |
| 0 | E102-2 | 42 | 11 | 0 | 5 | 0 | 4 | 2 | 20 |
| 0 | E103-2 | 14 | 2 | 0 | 3 | 1 | 1 | 0 | 7 |
| 4 | F101-0 | 25 | 20 | 0 | 1 | 0 | 0 | 3 | 1 |
| 2 | F103-0 | 104 | 64 | 4 | 12 | 14 | 0 | 9 | 1 |
| 2 | F104-0 | 102 | 67 | 9 | 16 | 1 | 1 | 6 | 2 |
| 2 | F105-0 | 82 | 44 | 6 | 20 | 6 | 0 | 6 | 0 |
| 2 | F106-1 | 53 | 37 | 1 | 8 | 1 | 1 | 4 | 1 |
| 2 | F107-1 | 27 | 15 | 3 | 5 | 1 | 0 | 3 | 0 |
| 2 | F103-3 | 162 | 103 | 12 | 6 | 30 | 1 | 9 | 1 |
| 2 | F104-3 | 154 | 86 | 5 | 36 | 12 | 2 | 13 | 0 |
| 2 | F105-3 | 23 | 11 | 1 | 7 | 3 | 0 | 1 | 0 |
| 2 | F108-4 | 42 | 30 | 3 | 2 | 6 | 0 | 1 | 0 |
| 2 | F105-5 | 52 | 27 | 0 | 18 | 3 | 0 | 4 | 0 |
| 2 | F106-5 | 28 | 9 | 4 | 13 | 2 | 0 | 0 | 0 |
| 2 | F102-6 | 156 | 112 | 4 | 6 | 25 | 1 | 5 | 3 |
| 2 | F103-6 | 202 | 113 | 27 | 28 | 6 | 0 | 27 | 1 |
| 1 | F104-6 | 38 | 26 | 3 | 6 | 0 | 0 | 1 | 2 |
| 2 | F206-1 | 17 | 9 | 0 | 1 | 5 | 1 | 1 | 0 |
| 4 | F301-0 | 47 | 23 | 7 | 4 | 8 | 0 | 1 | 4 |
| 3 | F302-0 | 37 | 14 | 6 | 2 | 10 | 0 | 5 | 0 |
| 3 | F303-0 | 77 | 30 | 5 | 13 | 24 | 0 | 5 | 0 |
| 2 | F304-0 | 26 | 10 | 10 | 1 | 3 | 0 | 2 | 0 |
| 2 | F306-0 | 23 | 14 | 7 | 1 | 1 | 0 | 0 | 0 |
| 1 | F307-0 | 23 | 11 | 9 | 1 | 0 | 0 | 2 | 0 |
| 1 | F308-0 | 38 | 16 | 8 | 9 | 1 | 0 | 4 | 0 |
| 3 | F302-1 | 46 | 20 | 5 | 0 | 16 | 0 | 4 | 1 |
| 2 | F303-1 | 14 | 8 | 1 | 0 | 4 | 1 | 0 | 0 |
| 2 | F302-2 | 13 | 8 | 0 | 0 | 4 | 1 | 0 | 0 |
| 2 | F304-2 | 26 | 12 | 5 | 8 | 1 | 0 | 0 | 0 |
| 3 | F303-3 | 48 | 23 | 1 | 0 | 24 | 0 | 0 | 0 |
| 2 | F304-3 | 64 | 43 | 6 | 2 | 6 | 0 | 7 | 0 |
| 2 | F305-3 | 40 | 19 | 7 | 4 | 1 | 0 | 6 | 3 |
| 1 | F306-3 | 37 | 13 | 15 | 6 | 0 | 0 | 3 | 0 |
| 1 | F307-3 | 42 | 19 | 15 | 2 | 0 | 0 | 6 | 0 |
| 1 | F308-3 | 10 | 7 | 0 | 2 | 0 | 0 | 1 | 0 |
| 1 | F407-0 | 23 | 15 | 2 | 4 | 0 | 1 | 1 | 0 |
| 3 | F503-0 | 29 | 16 | 1 | 1 | 6 | 2 | 3 | 0 |
| 3 | F504-0 | 55 | 29 | 1 | 0 | 24 | 1 | 0 | 0 |
| 2 | F505-0 | 127 | 88 | 5 | 3 | 29 | 0 | 2 | 0 |
| 2 | F506-0 | 81 | 66 | 3 | 3 | 7 | 0 | 2 | 0 |
| 2 | F507-0 | 98 | 59 | 7 | 17 | 8 | 1 | 6 | 0 |
| 1 | F509-0 | 61 | 37 | 1 | 19 | 1 | 0 | 3 | 0 |

| Phase | Level-Locus | Total | Fine | Sandy | Cream | Very Fine | Glaze | Hand-made | Coarse |
|---|---|---|---|---|---|---|---|---|---|
| 1 | F510-0 | 107 | 70 | 5 | 23 | 0 | 0 | 9 | 0 |
| 1 | F511-0 | 103 | 46 | 22 | 21 | 0 | 0 | 13 | 1 |
| 1 | F513-0 | 56 | 41 | 5 | 1 | 1 | 0 | 8 | 0 |
| 3 | F505-1 | 37 | 28 | 0 | 1 | 6 | 0 | 2 | 0 |
| 0 | F506-1 | 16 | 10 | 0 | 5 | 0 | 0 | 1 | 0 |
| 2 | F507-1 | 32 | 19 | 0 | 8 | 3 | 0 | 2 | 0 |
| 3 | F603-0 | 55 | 28 | 3 | 5 | 14 | 0 | 5 | 0 |
| 2 | F607-0 | 58 | 41 | 1 | 15 | 0 | 0 | 1 | 0 |
| 3 | F605-1 | 28 | 15 | 1 | 7 | 0 | 0 | 5 | 0 |
| 4 | G101-0 | 50 | 28 | 3 | 4 | 12 | 1 | 2 | 0 |
| 3 | G105-1 | 32 | 15 | 0 | 1 | 13 | 1 | 1 | 1 |
| 2 | G106-1 | 41 | 24 | 5 | 5 | 6 | 0 | 1 | 0 |
| 2 | G107-1 | 65 | 39 | 13 | 0 | 7 | 3 | 3 | 0 |
| 2 | G109-1 | 50 | 30 | 5 | 7 | 6 | 0 | 2 | 0 |
| 0 | G110-1 | 131 | 57 | 32 | 6 | 14 | 0 | 20 | 2 |
| 1 | G111-1 | 29 | 13 | 1 | 7 | 2 | 0 | 6 | 0 |
| 1 | G112-1 | 69 | 39 | 7 | 14 | 3 | 1 | 5 | 0 |
| 1 | G113-1 | 74 | 17 | 42 | 8 | 1 | 0 | 4 | 2 |
| 1 | G114-1 | 34 | 3 | 21 | 1 | 0 | 0 | 9 | 0 |
| 3 | G103-2 | 36 | 10 | 1 | 5 | 17 | 0 | 3 | 0 |
| 3 | G105-2 | 47 | 12 | 0 | 3 | 21 | 4 | 7 | 0 |
| 3 | G106-2 | 120 | 48 | 0 | 4 | 49 | 4 | 14 | 1 |
| 2 | G107-2 | 31 | 5 | 0 | 0 | 24 | 0 | 2 | 0 |
| 2 | G108-2 | 29 | 6 | 0 | 1 | 19 | 1 | 2 | 0 |
| 2 | G109-2 | 109 | 50 | 6 | 8 | 23 | 5 | 15 | 2 |
| 2 | G110-2 | 45 | 18 | 3 | 4 | 13 | 0 | 7 | 0 |
| 1 | G111-2 | 29 | 13 | 4 | 3 | 0 | 0 | 9 | 0 |
| 1 | G112-2 | 32 | 18 | 6 | 2 | 0 | 0 | 5 | 1 |
| 1 | G113-2 | 87 | 33 | 21 | 17 | 0 | 0 | 14 | 2 |
| 1 | G114-2 | 41 | 20 | 10 | 10 | 0 | 0 | 0 | 1 |
| 4 | G202-0 | 27 | 12 | 1 | 2 | 9 | 2 | 0 | 1 |
| 3 | G203-1 | 27 | 17 | 1 | 1 | 7 | 0 | 0 | 1 |
| 3 | G204-1 | 20 | 10 | 1 | 3 | 2 | 1 | 3 | 0 |
| 3 | G206-1 | 32 | 4 | 3 | 1 | 8 | 0 | 16 | 0 |
| 3 | G203-2 | 69 | 24 | 2 | 12 | 26 | 4 | 0 | 1 |
| 3 | G204-2 | 75 | 18 | 2 | 5 | 38 | 7 | 5 | 0 |
| 3 | G205-2 | 56 | 26 | 1 | 6 | 13 | 1 | 2 | 7 |
| 3 | G206-2 | 50 | 14 | 0 | 4 | 5 | 3 | 23 | 1 |
| 0 | G207-2 | 19 | 10 | 5 | 3 | 0 | 0 | 1 | 0 |
| 2 | G209-2 | 54 | 28 | 3 | 2 | 6 | 0 | 14 | 1 |
| 2 | G210-2 | 78 | 39 | 6 | 14 | 7 | 1 | 11 | 0 |
| 1 | G211-2 | 81 | 32 | 20 | 20 | 4 | 1 | 4 | 0 |
| 1 | G212-2 | 89 | 34 | 27 | 14 | 0 | 0 | 13 | 1 |
| 1 | G213-2 | 50 | 12 | 28 | 3 | 1 | 1 | 4 | 1 |
| 2 | G209-3 | 39 | 13 | 4 | 15 | 1 | 1 | 5 | 0 |
| 1 | G211-3 | 62 | 30 | 6 | 19 | 3 | 1 | 3 | 0 |
| 1 | G213-3 | 44 | 28 | 7 | 1 | 0 | 0 | 5 | 3 |
| 1 | G214-3 | 20 | 16 | 4 | 0 | 0 | 0 | 0 | 0 |
| 4 | G302-2 | 66 | 39 | 0 | 7 | 15 | 2 | 3 | 0 |
| 3 | G303-2 | 16 | 7 | 1 | 1 | 2 | 0 | 5 | 0 |
| 4 | G301-3 | 48 | 28 | 9 | 2 | 6 | 1 | 1 | 1 |

| Phase | Level-Locus | Total | Fine | Sandy | Cream | Very Fine | Glaze | Hand-made | Coarse |
|---|---|---|---|---|---|---|---|---|---|
| 4 | G302-3 | 64 | 38 | 7 | 7 | 8 | 2 | 1 | 1 |
| 3 | G303-3 | 61 | 25 | 1 | 6 | 26 | 0 | 3 | 0 |
| 4 | H101-0 | 23 | 17 | 2 | 2 | 2 | 0 | 0 | 0 |
| 0 | H104-0 | 12 | 8 | 0 | 1 | 0 | 0 | 0 | 3 |
| 0 | I204-2 | 20 | 5 | 0 | 0 | 0 | 3 | 1 | 11 |
| 0 | I208-0 | 13 | 8 | 0 | 1 | 0 | 0 | 2 | 2 |

APPENDIX 4

COUNTS OF MAJOR AL-BASRA POTTERY FORMS, BY PROVENIENCE UNITS (LEVEL-LOCUS)

| Phase | Level-Locus | Total | 40 | 41 | 42 | 43 | 44 | 48 | 93 | Other |
|---|---|---|---|---|---|---|---|---|---|---|
| 0 | A102-0 | 28 | 5 | 8 | 1 | 9 | 0 | 4 | 1 | 0 |
| 0 | C102-0 | 7 | 4 | 1 | 0 | 2 | 0 | 0 | 0 | 0 |
| 3 | D103-1 | 57 | 5 | 9 | 0 | 2 | 1 | 30 | 3 | 7 |
| 0 | E102-2 | 42 | 0 | 5 | 0 | 4 | 0 | 19 | 0 | 14 |
| 0 | E103-2 | 14 | 0 | 2 | 0 | 0 | 0 | 10 | 0 | 2 |
| 4 | F101-0 | 25 | 3 | 5 | 0 | 1 | 0 | 7 | 5 | 4 |
| 2 | F103-0 | 104 | 28 | 21 | 0 | 10 | 4 | 13 | 14 | 14 |
| 2 | F104-0 | 102 | 21 | 9 | 1 | 22 | 1 | 17 | 12 | 19 |
| 2 | F105-0 | 82 | 17 | 12 | 0 | 13 | 7 | 18 | 5 | 10 |
| 2 | F106-1 | 53 | 12 | 9 | 0 | 9 | 1 | 10 | 3 | 9 |
| 2 | F107-1 | 27 | 5 | 3 | 0 | 8 | 0 | 7 | 0 | 4 |
| 2 | F103-3 | 162 | 36 | 45 | 0 | 12 | 12 | 24 | 16 | 17 |
| 2 | F104-3 | 154 | 28 | 30 | 1 | 17 | 4 | 27 | 27 | 20 |
| 2 | F105-3 | 23 | 1 | 3 | 0 | 5 | 1 | 8 | 2 | 3 |
| 2 | F108-4 | 42 | 14 | 4 | 0 | 6 | 5 | 9 | 2 | 2 |
| 2 | F105-5 | 52 | 8 | 6 | 0 | 10 | 3 | 12 | 6 | 7 |
| 2 | F106-5 | 28 | 4 | 3 | 2 | 2 | 2 | 9 | 4 | 2 |
| 2 | F102-6 | 156 | 27 | 40 | 0 | 11 | 11 | 30 | 20 | 17 |
| 2 | F103-6 | 202 | 49 | 29 | 0 | 25 | 10 | 19 | 29 | 41 |
| 1 | F104-6 | 38 | 8 | 3 | 1 | 13 | 2 | 8 | 1 | 2 |
| 2 | F206-1 | 17 | 2 | 3 | 0 | 1 | 0 | 6 | 1 | 4 |
| 4 | F301-0 | 47 | 3 | 9 | 0 | 10 | 0 | 15 | 5 | 5 |
| 3 | F302-0 | 37 | 1 | 18 | 0 | 5 | 2 | 4 | 3 | 4 |
| 3 | F303-0 | 77 | 2 | 17 | 0 | 16 | 0 | 21 | 13 | 8 |
| 2 | F304-0 | 26 | 7 | 9 | 0 | 0 | 4 | 2 | 2 | 2 |
| 2 | F306-0 | 23 | 6 | 7 | 0 | 1 | 0 | 3 | 2 | 4 |
| 1 | F307-0 | 23 | 10 | 5 | 0 | 0 | 0 | 2 | 1 | 5 |
| 1 | F308-0 | 38 | 13 | 4 | 2 | 5 | 1 | 8 | 1 | 4 |
| 3 | F302-1 | 46 | 2 | 12 | 0 | 6 | 2 | 16 | 2 | 6 |
| 2 | F303-1 | 14 | 4 | 3 | 0 | 0 | 0 | 4 | 0 | 3 |
| 2 | F302-2 | 13 | 2 | 2 | 0 | 1 | 0 | 1 | 5 | 2 |
| 2 | F304-2 | 26 | 14 | 3 | 1 | 0 | 1 | 1 | 3 | 3 |
| 3 | F303-3 | 48 | 15 | 11 | 0 | 2 | 4 | 13 | 3 | 0 |
| 2 | F304-3 | 64 | 10 | 21 | 0 | 8 | 1 | 9 | 3 | 12 |
| 2 | F305-3 | 40 | 11 | 5 | 2 | 5 | 2 | 2 | 4 | 9 |
| 1 | F306-3 | 37 | 19 | 1 | 0 | 0 | 1 | 8 | 2 | 6 |
| 1 | F307-3 | 42 | 15 | 8 | 3 | 3 | 0 | 2 | 2 | 33 |
| 1 | F308-3 | 10 | 0 | 2 | 2 | 2 | 1 | 0 | 2 | 1 |
| 1 | F407-0 | 23 | 5 | 5 | 0 | 7 | 0 | 0 | 3 | 3 |
| 3 | F503-0 | 29 | 3 | 2 | 0 | 2 | 3 | 9 | 0 | 10 |
| 3 | F504-0 | 55 | 7 | 9 | 0 | 6 | 1 | 15 | 8 | 9 |
| 2 | F505-0 | 127 | 10 | 29 | 0 | 14 | 6 | 44 | 14 | 10 |
| 2 | F506-0 | 81 | 15 | 15 | 0 | 11 | 4 | 20 | 3 | 13 |
| 2 | F507-0 | 98 | 17 | 16 | 2 | 21 | 2 | 21 | 5 | 14 |
| 1 | F509-0 | 61 | 12 | 8 | 2 | 13 | 6 | 10 | 2 | 8 |

| Phase | Level-Locus | Total | 40 | 41 | 42 | 43 | 44 | 48 | 93 | Other |
|---|---|---|---|---|---|---|---|---|---|---|
| 1 | F510-0 | 107 | 23 | 9 | 4 | 31 | 3 | 19 | 4 | 14 |
| 1 | F511-0 | 103 | 14 | 10 | 12 | 29 | 0 | 11 | 4 | 23 |
| 1 | F513-0 | 56 | 9 | 5 | 7 | 15 | 2 | 5 | 1 | 12 |
| 3 | F505-1 | 37 | 11 | 3 | 0 | 7 | 2 | 9 | 2 | 4 |
| 0 | F506-1 | 16 | 1 | 1 | 0 | 2 | 0 | 7 | 4 | 1 |
| 2 | F507-1 | 32 | 11 | 4 | 0 | 2 | 0 | 11 | 1 | 3 |
| 3 | F603-0 | 55 | 5 | 14 | 0 | 2 | 2 | 12 | 7 | 13 |
| 2 | F607-0 | 58 | 11 | 17 | 0 | 10 | 6 | 5 | 5 | 4 |
| 3 | F605-1 | 28 | 5 | 4 | 0 | 5 | 1 | 3 | 4 | 6 |
| 4 | G101-0 | 50 | 4 | 10 | 0 | 5 | 0 | 16 | 9 | 6 |
| 3 | G105-1 | 32 | 2 | 8 | 1 | 4 | 0 | 7 | 6 | 4 |
| 2 | G106-1 | 41 | 7 | 8 | 0 | 5 | 3 | 9 | 3 | 6 |
| 2 | G107-1 | 65 | 17 | 6 | 4 | 11 | 0 | 14 | 6 | 7 |
| 2 | G109-1 | 50 | 6 | 9 | 0 | 4 | 1 | 13 | 9 | 8 |
| 0 | G110-1 | 131 | 13 | 34 | 1 | 16 | 9 | 18 | 11 | 29 |
| 1 | G111-1 | 29 | 4 | 4 | 0 | 3 | 3 | 6 | 3 | 6 |
| 1 | G112-1 | 69 | 19 | 10 | 1 | 17 | 1 | 7 | 6 | 8 |
| 1 | G113-1 | 74 | 6 | 3 | 10 | 30 | 2 | 6 | 1 | 16 |
| 1 | G114-1 | 34 | 2 | 0 | 15 | 2 | 0 | 1 | 0 | 14 |
| 3 | G103-2 | 36 | 2 | 7 | 0 | 2 | 2 | 13 | 3 | 7 |
| 3 | G105-2 | 47 | 6 | 6 | 0 | 2 | 1 | 9 | 6 | 17 |
| 3 | G106-2 | 120 | 11 | 25 | 0 | 6 | 7 | 31 | 10 | 30 |
| 2 | G107-2 | 31 | 12 | 6 | 0 | 1 | 3 | 7 | 0 | 2 |
| 2 | G108-2 | 29 | 3 | 5 | 0 | 5 | 2 | 6 | 2 | 6 |
| 2 | G109-2 | 109 | 19 | 17 | 0 | 6 | 8 | 18 | 9 | 32 |
| 2 | G110-2 | 45 | 5 | 14 | 0 | 4 | 3 | 7 | 1 | 11 |
| 1 | G111-2 | 29 | 4 | 2 | 0 | 6 | 0 | 2 | 4 | 11 |
| 1 | G112-2 | 32 | 6 | 3 | 0 | 12 | 3 | 0 | 0 | 8 |
| 1 | G113-2 | 87 | 11 | 6 | 0 | 34 | 4 | 13 | 1 | 18 |
| 1 | G114-2 | 41 | 4 | 9 | 6 | 13 | 0 | 5 | 0 | 4 |
| 4 | G202-0 | 27 | 2 | 9 | 0 | 2 | 1 | 13 | 0 | 0 |
| 3 | G203-1 | 27 | 1 | 3 | 0 | 1 | 3 | 16 | 1 | 2 |
| 3 | G204-1 | 20 | 2 | 2 | 0 | 0 | 1 | 8 | 0 | 7 |
| 3 | G206-1 | 32 | 0 | 0 | 0 | 0 | 0 | 13 | 1 | 18 |
| 3 | G203-2 | 69 | 2 | 17 | 0 | 7 | 2 | 32 | 3 | 6 |
| 3 | G204-2 | 75 | 7 | 7 | 0 | 8 | 5 | 30 | 5 | 13 |
| 3 | G205-2 | 56 | 6 | 6 | 0 | 15 | 2 | 17 | 0 | 10 |
| 3 | G206-2 | 50 | 3 | 4 | 0 | 3 | 4 | 5 | 1 | 30 |
| 0 | G207-2 | 19 | 1 | 4 | 2 | 2 | 0 | 6 | 1 | 3 |
| 2 | G209-2 | 54 | 11 | 10 | 1 | 4 | 0 | 7 | 0 | 21 |
| 2 | G210-2 | 78 | 14 | 11 | 0 | 12 | 2 | 14 | 7 | 18 |
| 1 | G211-2 | 81 | 13 | 6 | 2 | 33 | 3 | 11 | 2 | 11 |
| 1 | G212-2 | 89 | 8 | 5 | 15 | 30 | 1 | 9 | 0 | 21 |
| 1 | G213-2 | 50 | 5 | 6 | 16 | 12 | 0 | 0 | 4 | 7 |
| 2 | G209-3 | 39 | 4 | 8 | 0 | 6 | 3 | 4 | 5 | 9 |
| 1 | G211-3 | 62 | 9 | 6 | 1 | 21 | 5 | 5 | 5 | 10 |
| 1 | G213-3 | 44 | 8 | 0 | 10 | 7 | 1 | 1 | 3 | 14 |
| 1 | G214-3 | 20 | 4 | 2 | 2 | 7 | 2 | 3 | 0 | 0 |
| 4 | G302-2 | 66 | 5 | 14 | 1 | 4 | 4 | 21 | 7 | 10 |
| 3 | G303-2 | 16 | 2 | 3 | 0 | 0 | 1 | 3 | 2 | 5 |
| 4 | G301-3 | 48 | 9 | 19 | 0 | 2 | 4 | 8 | 0 | 6 |

| Phase | Level-Locus | Total | 40 | 41 | 42 | 43 | 44 | 48 | 93 | Other |
|---|---|---|---|---|---|---|---|---|---|---|
| 4 | G302-3 | 64 | 6 | 27 | 0 | 6 | 5 | 14 | 3 | 3 |
| 3 | G303-3 | 61 | 10 | 15 | 0 | 11 | 2 | 15 | 4 | 4 |
| 4 | H101-0 | 23 | 4 | 4 | 1 | 4 | 2 | 4 | 2 | 2 |
| 0 | H104-0 | 12 | 2 | 3 | 3 | 0 | 0 | 1 | 0 | 3 |
| 0 | I204-2 | 20 | 0 | 3 | 0 | 4 | 0 | 5 | 0 | 8 |
| 0 | I208-0 | 13 | 1 | 1 | 2 | 3 | 0 | 4 | 0 | 2 |

## APPENDIX 5

## THERMOLUMINESCENCE STUDY

A thermoluminescence (TL) study of fired clay material from al-Basra was undertaken in an attempt to get absolute dates for some of the deposits at the site. Although the historical record provides a few dates, e.g., its settlement in ca. A.D. 806, its destruction by the Umayyads in A.D. 979, and its final disappearance sometime in the eleventh or twelfth century, these do not help to age specific deposits, in particular the cultural material associated with each of the three building/occupation phases. In addition, the excavations failed to yield any evidence that could be used to date specific architectural features; the Islamic coins that were found were badly worn and have not be identified.

Conducted as a joint project between MASCA, University Museum, in Philadelphia, and the University of Durham in England, the TL study involved three samples: brick material from the F1 and F3 pottery kilns and burned red clay material from mud-brick walls whose collapse marked the end of the city. Only the F3 kiln sample produced a date, using a pre-dose technique (Fleming 1979): A.D. 825 ± 175 (Dur TL 34-4AS). Despite the disappointingly large standard deviation that represents close to 16% of the kiln's age, the date does place the kiln, which is stratigraphically associated with the earliest settlement at al-Basra, toward the first half of the city's history. The other two samples did not produce viable TL determinations. It is possible that the collapsed mud-brick walls did not heat up to the minimum temperatures needed for TL dating (ca. 450 degrees). It is unlikely, however, that the F1 kiln, which must have been heated repeatedly to temperatures ranging over 700 or 800 degrees centigrade, failed to reach the minimum. It seems that the broader question of TL dating of kilns may be fruitfully explored in the future, since very little work along these lines has been conducted thus far.

## BIBLIOGRAPHY


Abun-Nasr, Jamil M.

1975 *A History of the Maghrib.* 2nd edition. Cambridge University Press, Cambridge.

Adams, Robert McC.

1974 "Anthropological Perspectives on Ancient Trade." *Current Anthropology* 15: 239-258.

Adams, William Y.

1979 "On the Argument from Ceramics to History: A Challenge Based on Evidence from Medieval Nubia." *Current Anthropology* 20: 727-744.

Arnold, Dean E.

1975 "Ceramic Ecology of the Ayachucho Basin, Peru: Implications for Prehistory." *Current Anthropology* 16: 183-194.

1985 *Ceramic Theory: A Systems Approach to Cultural Process.* Cambridge University Press, Cambridge.

Ashtor, Eliyahu

1977 "Levantine Sugar Industry in the Later Middle Ages: An Example of Technological Decline." *Israel Oriental Studies* 7: 231-232.

Ataallah, M.

1967 "La Céramique musulmane à paroi fine incisée ou peinte de Lixus." *Bulletin d'Archéologie Marocaine* 7: 627-639.

al-Bakri, Abu-Ubayd (mid-11th century)

1965 *Description de l'Afrique Septentrionale.* Edited and translated by M. de Slane. 3d edition. Adrien-Maisonneuve, Paris.

Balfet, Hélène

1965 "Ethnographical Observations in North Africa and Archeological Interpretation." In *Ceramics and Man*, edited by Frederick Matson, pp. 161-177. Aldine Publishing Company, Chicago.

Bel, Alfred

1914 *Un Atelier de Poteries et Faiences du Xe Siècle de J.-C. Découvert à Tlemcen.* Braham, Constantine.

Benco, Nancy L.

n. d. "Chemical and Mineralogical Analyses of Early Islamic Pottery from al-Basra." Manuscript on file at Department of Anthropology, State University of New York, Binghamton.

1985 "Ceramic Diversity and Political Centralization: A Case Study from Medieval North Africa." Paper presented at the 50th Annual Meeting of the Society for American Archaeology, Denver, Colorado.

1986 "Diversity in Ceramic Production: A Case Study from Medieval North Africa." In Medieval Archaeology, edited by Charles L. Redman. Medieval and Renaissance Text Studies, State University of New York, Binghamton.

1986 "Organization and Technology of Pottery Production in Early Medieval Morocco." Ph.D. dissertation, State University of New York, Binghamton.

Boone, James L.

1980 "Artifact Deposition and Demographic Change: An Archeological Case Study of Medieval Colonialism in the Age of Expansion." Ph.D. dissertation, Department of Anthropology, State University of New York, Binghamton.

Bovill, E. W.

1968 The Golden Trade of the Moors. 2nd edition. Oxford University Press, London.

Bradley, Richard and Michael Fulford

1980 "Sherd Size in the Analysis of Occupation Debris." University of London, Institute of Archaeology, Bulletin 17: 85-94.

Brethes, J. D.

1939 Contribution à l'Histoire du Maroc par les Recherches Numismatiques. Imprimerie Les Annales Marocaines, Casablanca.

Brett, Michael

1972 "Problems in the Interpretation of the History of the Maghreb in the Light of Some Recent Publications." Journal of African History 13: 489-506.

Brewster-Wray, Tina

n. d. "Depositional Activities and Activity Sectors in a Huari Architectural Compound." Ms. on file, Department of Anthropology, State University of New York, Binghamton.

Brumfiel, Elizabeth M.

1980 "Specialization, Market Exchange and the Aztec State: A View from Huexotla." Current Anthropology 21: 459-479.

Bulliet, Richard W.

1975 The Camel and the Wheel. Harvard University Press, Cambridge, Massachusetts.

Canard, M.

1965 "Fatimids." In Encyclopaedia of Islam, New Edition, vol. II, pp. 850-862. E. J. Brill, Leiden.

Carcopino, Jerome

1945 Daily Life in Ancient Rome. Translated by E. O. Lorimer. Yale University Press, New Haven.

Charlton, Thomas

1978 "Teotihuacan, Tepeapulco and Obsidian Exploitation." Science 200: 1227-1236.

Childe, V. Gordon

1950 "The Urban Revolution." Town Planning Review 21: 3-17.

Conkey, Margaret

1980 "The Identification of Prehistoric Hunter-gatherer Sites: The Case of Altamira." Current Anthropology 21:609-630.

Daoulatli, A.

1979 Poteries et Céramiques Tunisiennes. Institut National d'Archéologie et d'Art, Tunis.

1980 "La céramique ifriqiyenne du IXe au XVe siècle." In La Céramique Médiévale en Méditerranée Occidentale Xe-XVe Siècle, edited by G. Démains d'Archimbaud and M. Picon, pp. 197-201. Éditions du Centre Nationale de la Recherche Scientifique, Paris.

Darling, M. J.

1977 "Pottery from Early Military Sites in Western Britain." In Roman Pottery Studies in Britain and Beyond, edited by J. Dore and K. Greene, pp. 57-100. International Series 30. British Archaeological Reports, Oxford.

DeBoer, Warren and James A. Moore

1982 "The Measurement and Meaning of Stylistic Diversity." Nawpa Pacha 20: 147-162.

Delphy, A.

1951-52 "Note sur une exposition temporaire de céramiques musulmanes archaiques trouvées au Maroc." Cahiers des Arts et Techniques de l'Afrique du Nord 1: 7-14.

1955 "Note sur quelques vestiges de céramique recueillis à Salé. Hespéris 42: 129-152.

Démains d'Archimbaud, G. and M. Picon (editors)

1980 La Céramique Médiévale en Méditerranée Occidentale: Xe-XVe siècles. Éditions du Centre National de la Recherche Scientifique, Paris.

Deverdun, Gaston and Marcel Rouche

1949 "Notes sur de nouveaux documents de céramique marocaine découverts à Marrakech." Hespéris 36: 451-455.

Dozy, Reinhart

1972 Spanish Islam: A History of the Moslems in Spain. Translated by Francis Griffin Stokes. Reprint of 1913 edition. Frank Cass, London.

Eustache, Daniel

1955 "El-Basra, Capitale Idrissite, et Son Port." Hespéris 42: 218-238.

1970-71 Corpus des Dirhams Idrisites et Contemporains. Banque du Maroc, Rabat.

1971a "Idris I." In Encyclopaedia of Islam, New Edition, Vol. III, p. 1031. E. J. Brill, Leiden.

1971b "Idris II." In Encyclopaedia of Islam, New Edition, Vol. III, pp. 1031-1032. E. J. Brill, Leiden.

1971c "Idrisids." In Encyclopaedia of Islam, New Edition, Vol. III, pp. 1035-1037. E. J. Brill, Leiden.

Ferron, J. and M. Pinard

1954 "Céramique musulmane à Carthage." Cahiers de Byrsa 4: 41-65.

Feinman, Gary

1980 "The Relationship between Administrative Organization and Ceramic Production in the Valley of Oaxaca." Ph.D. dissertation, City University of New York, New York.

1985 "Changes in the Organization of Ceramic Production in Pre-Hispanic Oaxaca, Mexico." In Decoding Prehistoric Ceramics, edited by Ben A. Nelson, pp. 195-223. Southern Illinois University Press, Carbondale.

Feinman, Gary, Stephen Kowalewski, Richard E. Blanton

1984 "Modelling Ceramic Production and Organizational Change in the Pre-Hispanic Valley of Oaxaca, Mexico." In The Many Dimensions of Pottery; Ceramics in Archaeology and Anthropology, edited by Sander van der Leeuw and Alison Pritchard, pp. 297-333. Amsterdam.

Feinman, Gary, Steadman Upham, Kent G. Lightfoot

1981 "The Production Step Measure: An Ordinal Measure of Labor Input in Ceramic Manufacture." American Antiquity 46: 871-884.

Février, P.A., A. Gaspary and R. Guéry

1970 Fouilles de Sétif (1959-1966). Supplement de Bulletin Archéoloque Algeria 1. Algiers.

Finley, M. I.

1971 "Archaeology and History." Daedalus 10: 168-186.

1973 The Ancient Economy. University of California Press, Berkeley.

Flannery, K. V.

1972 "The Cultural Evolution of Civilization." Annual Review of Ecology and Systematics 3:399-426.

Glick, Thomas F.

1979 Islamic and Christain Spain in the Early Middle Ages. Princeton University Press, Princeton, New Jersey.

Goitein, S. D.

1961 "The Main Industries of the Mediterranean Area as Reflected in the Records of the Cairo Geniza." Journal of Economic and Social History of the Orient 4: 168-197.

1967 A Mediterranean Society, vol. I: Economic Foundations. University of California Press, Berkeley.

1970 "Mediterranean Trade in the Eleventh Century: Some Facts and Problems." In Studies in the Economic History of the Middle East, edited by M. A. Cook, pp. 51-62. Oxford University Press, London.

1971 A Mediterranean Society, vol. 2: The Community. University of California Press, Berkeley.

1973 *Letters of Medieval Jewish Traders.* Princeton University Press, Princeton, New Jersey.

1983 *A Mediterranean Society, vol. 4: Daily Life.* University of California Press, Berkeley.

Golvin, Lucien

1965 *Recherches Archéologiques à la Qala des Banu Hammad (Algerie).* G. P. Maisonneuve et Larose, Paris.

1980 "Les Céramique émailées de période Hammâdide Qal'a des Banû Hammâd (Algéria)." In *La Céramique Médiévale en Méditerranée Occidentale: Xe-XVe siècles*, edited by G. Démains d'Archimbaud and M. Picon, pp. 203-217. Éditions du Centre Nationale de la Recherche Scientifique, Paris.

Grenier de Cardenal, Micheline

1980 "Recherches sur la céramique Médiévale Marocaine." In *La Céramique Médiévale en Méditerranée Occidentale: Xe-XVe siècles*, edited by G. Démains d'Archimbaud and M. Picon, pp. 227-250. Éditions du Centre Nationale de la Recherche Scientifique, Paris.

Hagstrum, Melissa B.

1985 "Measuring Prehistoric Ceramic Craft Specialization: A Test Case in the American Southwest." *Journal of Field Archaeology* 12: 65-75.

Harris, Edward C.

1975 "The Stratigraphic Sequence: A Question of Time." *World Archeology* 7: 109-121.

1979 *Principles of Archaeological Stratigraphy.* Academic Press, New York.

Hayes, John W.

1972 *Late Roman Pottery.* British School at Rome, London.

Haywood, R. M.

1956 "Roman Africa." In *An Economic Survey of Ancient Rome, vol. IV*, edited by Tenney Frank, pp. 1-120. Pageant Books, Inc., Paterson, New Jersey.

Hodder, Ian

1982 "Theoretical Archaeology: A Reactionary View." In *Symbolic and Structural Archaeology*, edited by Ian Hodder, pp. 1-16. Cambridge University Press, Cambridge.

Hole, Edwyn

1958 Andalus: Spain under the Muslims. London.

Hopkins, K.

1978 "Economic Growth and Towns in Classical Antiquity." In Towns in Societies, edited by P. Abrams and E. A. Wrigley, pp. 35-77. Cambridge University Press, Cambridge.

Ibn Abi Zar (d. 1310-20)

1860 Roudh el-Kartas, Histoire des Souverains du Maghreb, Espagne et Maroc, et Annales de la Ville de Fes. Translated by A. Beaumier. Paris.

Ibn Hawkal (ca. 977)

1842 "Description de l'Afrique." Translated from Arabic by M. de Slane. Journal Asiatique, 3rd series, vol. 13: 153ff.

Ibn Idhari (ca. 1306)

1901-1904 Histoire de l'Afrique et de l'Espagne. Translated by E. Fagnan. Alger.

Ibn Khaldun (ca. 1380)

1927 Histoire des Berberes. Vol. II. Translated by M. de Slane. Librairie Orientaliste, Paris.

al-Idrisi (1154)

1969 Description de l'Afrique et de l'Espagne. Edited and translated by R. Dozy and M. de Goeje. Reprint. Oriental Press, Amsterdam.

Imamuddin, S. M.

1965 Some Aspects of the Socio-Economic and Cultural History of Muslim Spain, 711-1492 A.D. E. J. Brill, Leiden.

Jeffers, Mary V.

1981 Domestic Architecture from Islamic Qsar es-Seghir. Master's thesis, Department of Anthropology, State University of New York, Binghamton.

Jenkins, Marilyn

1975 "Western Islamic Influences on Egyptian Iconography." Kunst des Orients 10:91-107.

1980 "Medieval Maghribi Luster-Painted Pottery." In La Céramique Médiévale en Méditerranée Occidentale: Xe-XVe siècles, edited by

G. Démains d'Archimbaud and M. Picon, pp. 335-342. Éditions du Centre Nationale de la Recherche Scientifique, Paris.

Johnson, Gregory A.

1973 *Local Exchange and Early State Development in Southwestern Iran*. Anthropological Papers No. 51. Museum of Anthropology, University of Michigan, Ann Arbor.

Jones, A. H. M.

1974 *The Roman Economy: Studies in Ancient Economic and Administrative History*. Rowman and Littlefield, Totowa, New Jersey.

Julien, Charles-André

1970 *History of North Africa: Tunisia, Algeria, Morocco.* Routledge and Kegan Paul, London.

Kintigh, Keith

1984a "Measuring Archaeological Diversity by Comparison with Simulated Assemblages." *American Antiquity* 49:44-54.

1984b "Sample Size, Significance, and Measures of Diversity." Paper presented at the 49th Annual Meeting of the Society for American Archaeology, Portland, Oregon.

1985 "Diversity Program V3.1--PC Version." Manuscript on file, 702 E. Victoria Street, Santa Barbara, California.

Kramers, J. H.

1936 "al-Mukaddasi." *Encyclopaedia of Islam. Vol. III*, pp. 708-709. E. J. Brill, Leiden.

Lacoste, Yves

1974 "The General Characteristics and Fundamental Structures of Medieval North African Society." *Economy and Society* 3: 1-17.

Lane, Arthur

1965 *Early Islamic Pottery; Mesopotamia, Egypt and Persia.* Faber and Faber, London.

Laroui, Abdallah

1977 *The History of the Maghreb: An Interpretive Essay.* Translated by Ralph Manheim. Princeton University Press, Princeton, New Jersey.

Lavoix, M. Henri

1891 Catalogue des Monnaies Musulmanes de la Bibliothèque Nationale: Espagne et Afrique. Imprimerie National, Paris.

Lawless, Richard I. and Gerald H. Blake

1976 Tlemcen: Continuity and Change in an Algerian Town. Westview Press, Boulder, Colorado.

Leo Africanus (16th century)

1956 Description de l'Afrique. Translated by A. Èpaulard. Adrien-Maisonneuve, Paris.

LeTourneau, Roger

1949 Fes Avant Le Protectorat. Casablanca.

1961 Fez in the Age of the Marinids. University of Oklahoma Press, Norman.

LeToureau, Roger and H. Terrasse

1965 "Fas." Encyclopaedia of Islam. New Edition. Vol. II, pp. 818-823. E. J. Brill, Leiden.

Lèvi-Provencal, E.

1944 Histoire de L'Espagne Musulmane. Vol. 1. L'Institut Francais d'Archèologie Orientale du Caire, Caire.

1960 "Abu 'Ubayd al-Bakri." Encylopaedia of Islam. New Edition. Vol. I, pp. 155-157.

Lèvi-Provencal, E. and L. Torres Balbas

1960 "Al-Andalus." Encylopaedia of Islam. New Edition. Vol. I, pp. 492-501. E. J. Brill, Leiden.

Lewis, Archibald R.

1951 Naval Power and Trade in the Mediterranean, A. D. 500-1100. Princeton, Princeton University Press.

Lewis, Bernard

1966 The Arabs in History. Harper and Row, New York.

Marcais, George

1913 Les Poteries et Faiences de la Qal'a des Beni Hammad (XIe siècle). Braham, Constantine.

1916 Les Poteries et Faiences de Bougie. Braham, Constantine.

1924 Recherches d'Archéologie Musulman en Tunisie. Bulletin de la Société Francaise des Fouilles Archéologiques, Paris.

1928 Les Faiences à Reflets Métalliques de la Grande Mosquée de Kairouan. Librairie Orientaliste Paul Geuthner, Paris.

1941 "La Berbérie au IXe siècle d'apres El-Ya'qoubi." Revue Africaine, pp. 40-61.

Marcais, George and A. Dessus-Lamare

1941 "Recherches d'archeologie musulmane Tihert-Tagdemt." Revue Africaine 90:24-57.

Marcais, George and J. Schacht

1960 "Aghlabids." Encyclopaedia of Islam. New Edition. vol. I, pp. 247-250. E. J. Brill, Leiden.

Marzouk, M. A.

1955 History of Textile Industry in Alexandria, 331 B.C.-1517 A.D. Alexandria University Press, Alexandria.

Meunié, Jacques, H. Terrasse, and G. Deverdun

1952 Recherches Archéologiques à Marrakech. Arts et Métiers Graphiques, Paris.

Mez, Adam

1937 The Renaissance of Islam. Translated by Salahuddin Khuda Baksh and D. S. Margoliouth. Jubilee Printing and Publishing House, Patna.

Mikesell, Marvin W.

1961 Northern Morocco: A Cultural Geography. University of California Press, Berkeley.

Miquel, A.

1971 "Ibn Hawqal." Encyclopaedia of Islam. New Edition. vol. III, pp. 787-788. E. J. Brill, Leiden.

Mohamedi, Anissa

1980 "Information sur les découvertes récentes en céramique médiévale effectuées à Sétif (Algérie)." In La Céramique Médiévale en Méditerranée Occidentale: Xe-XVe siècles, edited by G. Démains d'Archimbaud and M. Picon, pp. 219-220. Éditions du Centre Nationale de la Recherche Scientifique, Paris.

Morris, Craig

1974 "Reconstructing Patterns of Non-Agricultural Production in the Inca Economy: Archaeology and Documents in Institutional Analysis." In *Reconstructing Complex Societies*, edited by C. Moore, pp. 49-60. Bulletin of the American School of Oriental Research, Philadelphia.

1982 "The Infrastructure of Inka Control in the Peruvian Central Highlands." In *The Inca and Aztec States, 1400-1800; Anthropology and History*, edited by George A. Collier, Renato I. Rosaldo, John D. Wirth, pp. 153-171. Academic Press, New York.

Morris, Craig and D. E. Thompson

1970 "Huanuco Viejo: An Inca Administrative Center." *American Antiquity* 35: 344-362.

al-Mukaddasi (late 10th century)

1950 *Description de l'Occident Musulman au IVe-Xe siѐe* Edited and translated by C. Pellat. Editions Carbonel, Alger.

Myers, J. Emlen

1984 "The Political Economy of Ceramic Production: A Study of the Islamic Commonware Pottery of Medieval Qsar es-Seghir." Ph.D. dissertation, Department of Anthropology, State University of New York, Binghamton.

Nelson, Ben A. (editor)

1985 *Decoding Prehistoric Ceramics.* Southern Illinois University Press, Carbondale.

Nicklin, Keith

1971 "Stability and Innovation in Pottery Manufacture." *World Archaeology* 3:13-49.

Oman, G.

1971 "al-Idrisi." *Encyclopaedia of Islam. New Edition. vol. III*, pp. 1032-1035. E. J. Brill, Leiden.

Peacock, D. P. S.

1982 *Pottery in the Roman World: An Ethnoarchaeological Approach.* Longman, London.

Pianel, G.

1951 "La céramique de Négrine (IXe siѐcle)." *Hesperis* 38: 1-30.

Pielou, E. C.

1977 *Mathematical Ecology.* John Wiley and Sons, New York.

Posac Mon, Carlos

1960 "Datos Para La Archeologia Musulmana de Ceuta." *Hesperis-Tamuda* 1: 157-165.

1962 *Estudio Arqueologico de Ceuta.* Publicaciones del Instituto Nacional de Ensenanza Media, Ceuta.

Rathje, William L.

1971 "The Origin and Development of Lowland Classic Maya Civilization." *American Antiquity* 36:275-285.

1975 "The Last Tango in Mayapan: A Tentative Trajectory of Production-Distribution Systems." In *Trade and Exchange in Ancient Civilization*, edited by J. Sabloff and C. C. Lamberg-Karlovsky, pp. 409-448. University of New Mexico Press, Albuquerque.

Redman, Charles L.

1980 "Late Medieval Ceramics from Qsar es-Seghir." In *La Céramique Médiévale en Méditerranée Occidentale: Xe-XVe siècles*, edited by G. Démains d'Archimbaud and M. Picon, pp. 251-263. Éditions du Centre Nationale de la Recherche Scientifique, Paris.

1983 "Comparative Urbanism in the Islamic Far West." *World Archaeology* 14: 355-377.

1985 *Qsar es-Seghir: An Archaeological View of Medieval Life.* Academic Press, New York.

in press "Survey and Test Excavation of Six Medieval Islamic Sites in Northern Morocco." *Bulletin d'Archéologie Marocaine* 14.

Redman, Charles L., M. Jeffers, A. Magennis, J. E. Myers, S. Pollock, A. Wondriska

1980 "Progress Report of the Moroccan-American Archeological Project: Number 12, Qsar es-Seghir, Al-Basra, Badis, and Nakur." August 10, 1980 to October 5, 1980.

Renfrew, Colin

1979 "Introduction: Production and Exchange in Early State Societies, the Evidence of Pottery." In *Pottery and Early Commerce,* edited by D. P. S. Peacock, pp. 1-20. Academic Press, New York.

Rhodes, Daniel

1981 Kilns: Design, Construction and Operation. Chilton Book Company, Philadelphia.

Ricard, Prosper and Alexandre Delphy

1931 "Notes sur la dècouverte de spècimens de cèramique marocaine du Moyen-Age." Hesperis 13: 227-270.

Rice, Prudence M.

1977 "Whiteware Pottery Production in the Valley of Guatemala: Specialization and Resource Utilization." Journal of Field Archaeology 4:221-233.

1981 "Evolution of Specialized Pottery Production: A Trial Model." Current Anthropology 22: 219-240.

Riley, John

1979-80 "Industrial Standardization in Cyrenaica during the Second and Third Centuries A.D.: The Evidence from Locally-Manufactured Pottery." Society for Libyan Studies 11th Annual Report (1979-80), pp. 73-78.

Rottlander, R. C. A.

1966 "Is Provincial Roman Pottery Standardized." Archaeometry 9: 76-92.

Rubertone, Patricia

1979 "Social Organization in an Islamic Town: The Behavioral Explanation of Ceramic Variability." Ph.D. dissertation, Department of Anthropology, State University of New York, Binghamton.

Rye, Owen S.

1981 Pottery Technology: Principles and Reconstruction. Taraxacum Inc., Washington, D.C.

Rye, Owen S. and Clifford Evans

1976 Traditional Pottery Techniques of Pakistan: Field and Laboratory Studies. Smithsonian Contributions to Anthropology 21, Smithsonian Institution Press, Washington, D.C.

Sanders, W. T.

1968 "Hydraulic Agriculture, Economic Symbiosis, and the Evolution of States in Central Mexico." In Rise and Fall of Civilizations, edited by Jeremy Sabloff and C. C. Lamberg-Karlovsky, pp. 119-133. Cummings Publishing Company, Menlo Park, California.

Schiffer, Michael B.

1976 Behavioral Archaeology. Academic Press, New York.

1983 "Toward the Identification of Formation Processes." American Antiquity 48: 675-706.

Schulman, Jay

n. d. "The History of Qsar es-Seghir." Manuscript on file, Department of Anthropology, Arizona State University, Tempe.

Shepard, Anna O.

1956 Ceramics for the Archaeologist. Carnegie Institution of Washington Publication 609, Washington, D.C.

1963 "Beginnings of Ceramic Industrialization: An Example from the Oaxaca Valley." Notes from a Ceramic Laboratory No. 2. Carnegie Institution of Washington, Washington, D.C.

Skinner, G. William

1977a "Regional Urbanization in Nineteenth-Century China." In The City in Late Imperial China, edited by G. W. Skinner, pp. 211-249. Stanford University Press, Stanford.

1977b "Cities and the Hierarchy of Local Systems." In The City in Lake Imperial China, edited by G. W. Skinner, pp. 275-351. Stanford University Press, Stanford.

Smith, Carol A.

1976 "Regional Economic Systems: Linking Geographical Models and Socioeconomic Problems." In Regional Analysis. vol. 1: Economic Systems, edited by C. Smith, pp. 3-63.

Spence, Michael W.

1981 "Obsidian Production and the State in Teotihuacan." American Antiquity 46: 769-788.

Steponaitis, Vincas P.

1983 Ceramics, Chronology, and Community Patterns: A Methodological Study at Moundville. Academic Press, New York.

Terrasse, Henri

1937 "La céramique Hispano-Maghrébine du XIIe siècle d'après les fouilles du Château de l'Ain Ghaboula (Dchîra)." Hesperis 24: 13-18.

1949-1950 Histoire du Maroc des Origines à l'Etablissement du Protectorate Francais. 2 volumes. Casablanca.

Terrasse, Henri and Jacques Meunier

1957 Nouvelles Recherches Archèologiques à Marrakech. Publications de l'Institut des Hautes Etudes Marocaines 62, Paris.

Terrasse, Michel

1976 "Recherches archèologiques d'èpoque islamique en Afrique du Nord." Acadèmie des Inscriptions et Belles-Lettres, Comptes Rendus, November-December 1976:590-611.

Tissot, M. Charles

1877 Recherches sur la Gèographie Comparèe de la Maurètanie Tingitane. Paris.

Unger, Richard W.

1980 The Ship in the Medieval Economy, 600-1600. Croon Helm, London.

Van der Leeuw, Sander

1976 Studies in the Technology of Ancient Pottery. Amsterdam.

Van der Leeuw, Sander and Alison C. Pritchard (editors)

1984 The Many Dimensions of Pottery, Ceramics in Archaeology and Anthropology. Amsterdam.

Van Gennep, M.

1911 Etudes d'Ethnographie Algerienne Rèvue d'Ethnographie et de Sociologie. E. Leroux, Paris.

Velàzquez Bosco, R.

1912 Medina Azzahra y Alamiriya. Madrid.

1923 "Excavaciones en Medina Azahara." MJSEAA 54.

Vitelli, Giovanna

1981 Islamic Carthage: The Archaeological, Historical, and Ceramic Evidence. Centre d'Etudes et de Documentation Archèologique de Carthage (CEDAC), Institut National d'Archèologie et d'Art, Carthage.

Vitelli, Giovanna and D. Pringle

1978 "Bibliography of North African Medieval Pottery." Medieval Ceramics 2: 53-58.

Vossen, Rudiger

1984 "Towards Building Models of Traditional Trade in Ceramics: Case Studies from Spain and Morocco." In The Many Dimensions of Pottery, Ceramics in Archaeology and Anthropology edited by S. van der Leeuw and A. Pritchard, pp. 341-406. Amsterdam.

Vossen, Rudiger and Wilhelm Ebert

1986 Marokkanische Topferei: Topferorte und -zentren Eine Landesaufnahme (1980). Dr. Rudolf Habelt GmbH, Bonn.

Waltzing, J. P.

1968 Étude Historique sur les Corporations Professionnelles Chez les Romains. Reprint of 1900 edition. Forni Editore, Bologna.

Watt, W. Montgomery

1965 A History of Islamic Spain. Edinburgh University Press, Edinburgh.

Wattenmaker, Patricia

1981 "Progress Report on Animal Bones from Badis, al-Basra, Jebila, Moulay bou-Selham, Nakur and Qsar es-Seghir. Manuscript on file, Department of Anthropology, Arizona State University, Tempe.

Whitehouse, David

1970 "Siraf: A Medieval Port on the Persian Gulf." World Archaeology 2: 141-158.

Wilkinson, C. K.

1959 "The Kilns of Nishapur." Bulletin. Metropolitan Museum of Art:

Williams, D. F.

1977 "The Romano-British Black-Burnished Industry: An Essay on Characterisation by Heavy Mineral Analysis." In Pottery and Early Commerce: Characterisation and Trade in Roman and Later Ceramics, edited by D. P. S. Peacock, pp. 163-220. Academic Press, London.

Wittfogel, Karl

1957 Oriental Despotism. Yale University Press, New Haven.

Wright, Henry T.

1969 The Administration of Rural Production in an Early Mesopotamian Town. Anthropological Papers No. 51. Museum of

Anthropology, University of Michigan, Ann Arbor.

1981 An Early Town on the Deh Luran Plain: Excavations at Tepe Farukhabad. Memoirs No. 13. Museum of Anthropology, University of Michigan, Ann Arbor.

Wright, Henry T. and Gregory Johnson

1975 "Population, Exchange, and Early State Formation in Southwestern Iran." American Anthropologist 77: 267-289.

Wulff, Hans E.

1966 The Traditional Crafts of Persia. M.I.T. Press, Cambridge, Massachusetts.

Young, Christopher

1977 "Oxford Ware and the Roman Army." In Roman Pottery Studies in Britain and Beyond, edited by J. Dore and K. Greene, pp. 289-293. International Series 30. British Archaeological Reports, Oxford.

Yver, G.

1960 "al-Basra." Encyclopaedia of Islam. New Edition. vol.1, pp. 1087-1088. E. J. Brill, Leiden.

Zbiss, S. M.

1956 "Mahdia et Sabra-Mansoûriya, Nouveaux Documents d'Art Fatimite d'Occident." Journal Asiatique 244: 79-93.

Zozoya, Juan

1969 "Red-Painted and Glazed Pottery in Western Europe: Spain." Medieval Archaeology 13: 133-136.

1980a "Apercu général sur la céramique espanole." In La Céramique Médiévale en Méditerranée Occidentale: Xe-XVe siècles, edited by G. Démains d'Archimbaud and M. Picon, pp. 265-296. Éditions du Centre Nationale de la Recherche Scientifique, Paris.

1980b "Essai de chronologie pour certain types de céramique califale andalouse." In La Céramique Médiévale en Méditerranée Occidentale: Xe-XVe siècles, edited by G. Démains d'Archimbaud and M. Picon, pp. 311-315. Éditions du Centre Nationale de la Recherche Scientifique, Paris.

www.ingramcontent.com/pod-product-compliance
Lightning Source LLC
LaVergne TN
LVHW071207210926
848482LV00015B/746
* 9 7 8 0 8 6 0 5 4 4 4 0 1 *